The Flier's World

The Fliers' World

By James Gilbert

A Ridge Press Book | Grosset & Dunlap

Editor-in-Chief: Jerry Mason
Editor: Adolph Suehsdorf
Art Director: Albert Squillace
Associate Editor: Ronne Peltzman
Associate Editor: Joan Fisher
Art Associate: David Namias
Art Associate: Nancy Louie
Art Production: Doris Mullane
Picture Research: Marion Geisinger

Project Art Director: Harry Brocke
Associate Art Director: Tony Pollicino

Portions of Chapter 5 originally appeared
in *Flying* magazine and are reprinted courtesy of
Ziff-Davis Publishing Company.

Library of Congress Cataloging in Publication Data
Gilbert, James
        The flier's world.
1. Aeronautics—Popular works. I. Title.
TL546.7.G53   629.13   76-12562

Printed in the Netherlands by Smeets Offset, Weert.

To my fellow members of the Tiger Club
at Redhill in England,
in whose company my own enthusiasm for
flying first flourished.

Contents

Introduction

Cocktail-party chatter: "So you fly, do you? How interesting." There's usually a short pause, then the questions—always the same questions.

"Been at it long?" (As though it were a temporary aberration that ought soon to pass.) Since I was a kid of seventeen.

Another short pause while the astounding news that kids of seventeen are allowed to fly is digested. Then:

"Where do you fly?" (This usually from women, who tend to view flying as one of those things, like golf, that men go to on Sunday mornings, just to get out of the house.) Anywhere, ma'am. This year I have flown in and out of twenty-five airports and two cow pastures, one of them remarkably stony.

"Do you . . . ?" Own my own airplane? Yes. It's an antique aerobatic biplane, open cockpit, and it's nearly as old as I am.

"Oh, is it . . . ?" Safe? "I mean, being that old and everything." It is as robust as the day they first pushed it out of the factory in Switzerland, and once a year it is carefully inspected by skilled mechanics (as it must be by law) to be sure it has developed no defects. And it lives, when not flying, in a dry hangar—unlike my automobile, which sits out in the weather.

"Have you ever . . . ?" Had a crash? Nope. Nor even been close. Cracked up my car once, and fell off a ladder painting the porch, but have had no problems in the air. The most dangerous part of flying is probably—surprise—driving to the airport.

"Can you . . . ?" Fly jets? Well, to fly any of the jets I've ever met, you need a special license for that particular type, as well as a simple pilot's license and, no, I don't hold any jet-type ratings. But I have on one or two happy occasions flown a jet from the copilot's seat, under the supervision of a properly qualified pilot as captain.

"Are you allowed to . . . ?" Carry passengers? Yes. Once you have gained a Private Pilot's License, which normally takes thirty-five or forty hours of flying. Until then you may fly only with an instructor, or solo only when he says so. To carry passengers and be paid for it you need rather more experience and a commercial pilot's license.

"Isn't it . . . ?" Very expensive? Well, yes, it is expensive. That's the one big snag. But life is expensive and getting more so all the time. (Do golfers, for instance, ever truly figure the cost of golfing? The club subscriptions, the new set of clubs, greens fees at other courses, the round of drinks in the bar afterward?) I reckon my airplane costs *about thirty-five dollars an hour to run, all in, and an hour a week I can easily afford. You can learn to fly, from scratch to a PPL, for well under a thousand bucks, which these days is no more than it costs to trade in your old car for a newer model. Keep your old one for an extra year or two, and you can learn to fly on what you save.*

"I've always thought I'd love to try . . ." Gliding? There are gliding schools, too, usually separate from power-flying schools. It is a simpler, freer type of flying, and cheaper as well. The new sport of hang-gliding is the cheapest, simplest way aloft of all. You don't even need a license or instruction to try it, though lessons are available if you want them.

"Can anyone . . . ?" Learn to fly? Yes, just about. You do have to pass a simple physical, but anyone who's reasonably fit will qualify, spectacle-wearers and all. Nor do you have to be very young. At least one seventy-year-old has learned to fly, and many people successfully take up flying in middle age, after their kids are through school and they at last have the spare cash for such an adventure.

"Do many . . . ?" Women fly? Yes, an increasing number. As a matter of fact, my first instructor, who took me through to my PPL, was a woman.

"How do I . . . ?" Start? Pick a flying school and ask for a trial lesson. If you enjoy it and decide to continue, they'll tell you what to do next, and where to go for your physical.

It is to answer questions like these, from people half-fascinated by the world of aviation but baffled about how it works, that I have written this book. I hope it may serve to encourage some of them to take up flying, as well as to encourage some who are already fliers to widen and deepen their involvement in flying, for this book is by no means only for newcomers to aviation.

Welcome, then, to the flier's world.

✳   ✳   ✳

I'd like to thank those of my friends and colleagues in aviation who helped me with this book: Alan Bramson, Captain David Gibson of the 20th Tactical Fighter Wing, USAF, Ed Mack Miller of United Airlines, Richard L. Collins of Flying *magazine, Don Berliner, Robert Buck, Peter Garrison, and George Moffat, who read and checked chapters of the manuscript for me; Robert B. Parke (editor and publisher of* Flying*), David Esler, David Namias, and Berl Brechner, who supplied photographs; and Ernest Gann, who encouraged me strongly during the book's preparation.*

—James Gilbert

First Solo

You and your instructor have been going round and round the airport in the little trainer for the best part of an hour, though it feels more like all morning. The exercise you've been practicing is "shooting touch-and-goes"—approach the airport, touch down at the start of the runway as though you were landing, then open the throttle and climb away and around and back again. He misses nothing, your instructor: the airspeed shown on your dial should be locked on 90 mph (145 km/h) as you climb away; if it climbs to 100 or falls below 85 while you're looking around for other traffic, he'll draw your attention to it. And he won't let you get away with 900 feet (274 m) of altitude, or 1,100 on the downwind leg. He wants it exactly 1,000 feet on the altimeter, as the used-car lots, the hamburger heavens, and the new thruway slide by below you.

This is perhaps your tenth hour of training. You may be only seventeen, still a kid in your parents' eyes, blowing the savings from your summer job on the one thing you've dreamed of doing for years: learning to fly. You may be in your forties, with your own kids starting to earn for themselves, leaving you with some cash you didn't know what to do with, only you did know at heart. You've known for years what you wanted to do when you got the chance: learn to fly. You might even be in your sixties, just retired, following your family's fond advice to get out of the house and find a new interest. What an interest! Perhaps you're a woman; plenty of women fly. Or your instructor might be a woman; mine was. The chances are, though, that your instructor is a young man "building hours"—gaining experience in small aircraft until he begins to look attractive to the pilot-hiring office of one of the airlines. Still, he may be an excellent instructor, even though his career plans are elsewhere. The airlines are where the real money is for pilots.

But flight instructors are a various lot. The flight school you're with now may not be the one you started with. You may have picked a bad one at first, and after putting up with poor instruction, broken appointments, and a general lack of enthusiasm, you may have transferred to the school across the field. A good flight school should have bright new training aircraft on the ramp, tidy offices, class-rooms or a lounge for students (with clean ashtrays and a beverage vending machine that works), airplanes that really are available at the hour you reserve them for, and instructors who take time before or after flying to explain what they are trying to teach you in the air. Ground school is an important part of flying training. There's some inevitable book work to be done on navigation, weather, theory of flight, and so on. Most flight schools are good and getting better, a joy to do business with. Considering what flying costs, they should be!

No doubt you checked all this out when you took your first lesson—a trial lesson, probably, maybe just five dollars' worth, only a half-hour in the air, to see if you liked flying. You did like it, of course, and when the instructor showed you how the controls worked, and explained the instruments on the panel, you may well have been surprised at how simple an airplane is to control—at least in straight and level flight.

The main control is the yoke. This is usually a double-handled thing shaped not unlike a ram's horns, projecting out toward you from the bottom of the instrument panel. The yoke takes many different forms. In some airplanes, particularly sports planes, it may even be a floor-mounted stick. Whatever its shape, it always works in the same way: pull it toward you and it raises hinged surfaces (the elevators) on the tail plane, which raise the aircraft's nose in flight to make it climb. Ease forward and the nose lowers, causing you to descend. Tilt it left and right and it raises and lowers another set of hinged surfaces, called ailerons, on the wings. These move differentially (if the left ailerons are up, the right ones are down) to make the airplane roll left or right—which it needs to do to turn, since airplanes can only make banked turns. Use of the yoke comes to you very quickly, once you realize that its operation is perfectly natural. Any way you move the yoke the aircraft will tilt to follow it.

The rudder is worked differently—by floor pedals. The rudder is the least-used of an aircraft's aerodynamic controls. You need it to stay accurately aligned with the centerline of the runway while approaching to land and while climbing away after takeoff; and you use it in coordination with the eleva-

Preceding pages: Student pilot practices
"shooting touch-and-goes" with instructor in Beechcraft
Sundowner 180. He is approaching runway,
will touch down as though landing, then open throttle,
climb away, and repeat—while heeding airspeed, altitude,
other traffic, and instructor's instructions.

tor and ailerons to make your turns accurate, maintaining a level altitude with no yaw (skidding left or right). You may also need to use the rudder in slow flight, when the aircraft becomes less stable and is inclined to yaw.

Engine power is controlled not with a foot pedal, as in an automobile, but by a hand throttle, either a plunger type or a hinged-quadrant type, like a paper guillotine. Forward for full power, back toward you for idle.

Aero engines also have a fuel-air mixture control to allow you to meter the amount of fuel being fed to match reduced air pressure at altitude. And there is usually a carburetor heat control, used to prevent any build-up of ice in the carburetor when the plane is flying at reduced power. The starter and ignition switch is usually like that in a car, except that it has five positions—off, 1, 2, 1 + 2, and start —because an aircraft engine has dual ignition systems for safety.

There's a hand brake for the aircraft's main wheels, and the rudder pedals often have toe brakes as well. There is a lever to raise and lower the flaps at the rear of the wings that give extra lift at slow landing speeds. And there is a knurled wheel, called the trimmer, which enables you to compensate for the varying loads on the elevators (they change with airspeed and power changes), so that you don't have to exert a steady pull or push on the yoke to stay in level flight.

It is the number of instruments in even a simple training plane that most impresses the novice. But they quickly become familiar. The workings of some are obvious indeed: airspeed indicator, altimeter, clock, vertical-speed indicator (climb or descent), engine tachometer, fuel gauges, and so on. There is also the turn coordinator, a gyroscopic instrument that shows how steeply you are turning and whether you are slipping or yawing. The attitude indicator, another gyro instrument, shows a little model plane against an earth-and-sky background, which moves to show you where the horizon is in relation to your full-size airplane. The direction indicator shows your compass heading. And to round off the panel, there are radio sets for communication and navigation; a neat row of switches for the air-

craft lights and other electrical equipment; and a row of replacement fuses.

On that first instructional flight your instructor might not have explained things in as much detail as I have here. Probably he simply let you try the controls, and maybe he only referred you to the instruments to show you how you were doing.

And, of course, after you landed there was never any doubt that you would go on with learning to fly. The next step? To make another appointment with the flight school—and one with a local doctor who was also a Federal Aviation Authority-approved medical examiner. He gave you a simple physical, and there and then he issued you a student pilot's permit. When you went for your next lesson at the airport they sold you a blue pilot's logbook from the glass sales cabinet filled with instruction books and navigation instruments. When you wrote your name in the front, you knew you were on your way.

In your first hours in the air you were taught straight and level flight, and to glance at the altimeter and direction indicator to confirm that straight is straight, and level level. You weren't to keep your eyes down for long, though. You had to keep a constant lookout for other aerial traffic. You were taught how to turn (and maintain altitude while turning), how to glide (throttled back, remembering to pull the carburetor heat knob first), and how to climb at a steady angle and airspeed. And then your instructor made you put these together and turn while descending or climbing. At a safe altitude he showed you stalls—what happens if you slow up so much that the orderly airflow over the wings breaks down. The stall warning buzzer sounds, and the nose drops of its own accord as the aircraft struggles to regain flying speed. You practiced departure stalls, with full power, off a climbing turn; and level stalls, with the wing flaps down—always with that damn buzzer going. There's a lot of this in aviation: deliberately practicing something that you expect never to encounter, just in case. Almost everybody learning to fly is surprised to find that the training (which follows a Federal government syllabus) is far more thorough than expected.

And you practiced figure-eight maneuvers, to promote a smooth touch on the controls and coordin-

13

In the flier's world, most planes are little,
most pilots on the first rung of aviation achievement.
But this is quite enough for fun in the air,
whether landing in New Hampshire, spinning a prop in
Iowa, taxiing at sundown in Utah,
or otherwise coming, going, tinkering, or
relaxing in the shade of a stabilizer.

ation between hands and feet, eyes and brain. Then flights like today's, where you never left the traffic pattern and just practiced takeoffs and landings. Learning to land is the hardest task for a student: learning to keep your airspeed constant on final approach, reaching out with your eyes for the approaching runway and learning to judge the exact moment to close the throttle and pull back on the control yoke gently, gently, till your wheels kiss the concrete with a little *chirrup*, just as your flying speed is all but gone. At first, perhaps, you wondered if you'd ever get it right. Sometimes you "rounded out" too high and sank to the runway with an ungainly bump, or you touched down too fast, with a ten-foot bounce that had your instructor reaching for the throttle to apply power so you could go round and try it again. At the same time, you were learning to use the aircraft's two-way radio, slowly and hesitantly and sometimes confused by the jargon at first, but quickly becoming familiar with it. "Downwind leg" is when you turn parallel to the runway you took off from and head back for another landing; "base" is the segment of the nearly square traffic pattern you fly at right angles to the runway just before turning onto "finals"—your final approach and touchdown. The airport tower controller will instruct you to "rock your wings" if he's not sure which one of two similar aircraft in the pattern you are. "Go around," he orders if you get too close to the aircraft landing ahead of you. "Caution wake turbulence departing traffic," he'll warn if a big aircraft should take off ahead of you. His instructions are always preceded by your call-sign, which is the registration number painted on the side of your fuselage and written out on the instrument panel: N2153X, say. The N indicates that the aircraft is of United States registry, and is usually omitted in radio calls, so that you become "Two One Five Three X-ray," often abbreviated to "Five Three X" (formally, though, you should use the phonetic alphabet to indicate your final letter: Alpha, Bravo, Charlie, Delta, and so on). "Five Three X-ray, downwind, turning base," you'll call, and the tower will come back: "Five Three X, number three, report finals." Meaning there are two aircraft ahead of you, make sure you have them in sight, and then call again on

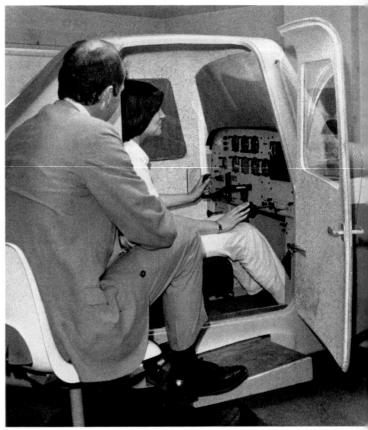

final approach. And be sure he has said "Cleared to land" before your wheels touch the runway.

By now you have discovered that set routines and procedures are a big part of flying. Before you even climbed aboard your trainer you learned to "preflight" it—to check the oil level in the engine and the fuel levels in the tanks (you never totally trust the electric gauges in the cockpit), to walk around the airplane, examining it minutely to make certain it is in flying condition. You even drain a little gas from it to be sure there is no water contamination. There are takeoff checks to be run through before you taxi out onto the runway, and a similar routine before landing. "Check lists" are printed in the aircraft's flight manual and on the cockpit panel, and your instructor makes you read these aloud, to help you get into the habit of following them religiously.

Process of Learning: A navigating problem and student in simulator (opposite), preflight checks of stabilator (below l) and aileron (bottom). Details: Stall-warning tab, purging water from fuel tank, Pitot tube (airspeed indicator), disc brake.

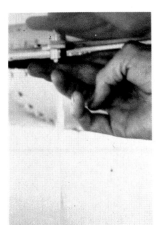

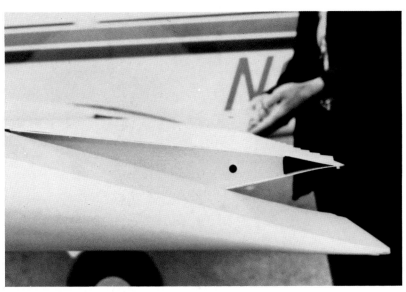

Coming around for next touch-and-go (l),
pilot banks, then levels plane (above & below)
for touch-down. Note turn instrument's
faithful indication (over pilot's right shoulder)
of Sundowner's relation to horizon.

19

So there you are, this fine morning, shooting touch-and-goes and feeling that one day soon you might get the hang of it. Not that your landings are perfect "greasers"—so smooth the runway seems greased—but they're not real "bouncers" either. Then, as you are established on finals, your instructor reaches across, picks up the microphone, and says to the tower, "Five Three X, full stop this time." You're somewhat annoyed that he seems to have had enough of you for the day, for you still have twenty minutes of your appointment left. But after you've landed and taxied off the runway, he says, all casual like, "Why don't you taxi back to the school and drop me off and then do one on your own?"

Your first solo! If there's powerful excitement in the idea for you, there seems to be none for him. "You'll find she climbs faster without my weight," he tells you, adding instructions to watch your speed on finals, to keep a good lookout for other traffic, and not to be ashamed to go around if you bounce the first time. "Not that you will," he says. "First one you do on your own is usually the best landing you'll ever do in your whole life." In fact, he says, if you feel happy about it, you can do three touch-and-goes. But come back in if the tower tells you to. And he opens the door and climbs out, pausing only to do up the loose ends of his seat strap. Without so much as a backward glance, he closes the airplane door and heads for the flight office.

As you taxi back to the runway, you can't help feeling a strange elation beneath your surface calm. For if the moment when you wrote your name in your own pilot's logbook was the moment you first knew you were on your way, then your first solo is your first and most marvelous achievement as a pilot. And your instructor was right, she does climb a little faster without him; without him watching you seem to fly better, too.

You are permitted to sing to yourself on downwind leg on your first solo—at least, I did. (But remember your instructor's warning to watch for other traffic.) Base, and then finals; carburetor heat, electric boost pump, and flaps—it has never seemed this easy. "Five Three X, finals, touch and go," you tell the tower. He knows, though, for your instructor is up in his cabin with him, not quite so cool about events now that you can't see him. After touch-down, as you push in the carburetor heat and apply power to take off again, you realize with awe that he was right, that *was* the best landing you've ever done, a perfect greaser. The second one, however, doesn't work out quite so well, and you bang the wheels onto the ground a little untidily. The third one is better, not an absolute greaser, but very fair. As you taxi back to the flight office your instructor comes out to meet you. "How did it go, then?" he asks with a smile, as if he hadn't been watching you, hawk-eyed, for every minute of those three trips around the traffic pattern. "Congratulations," he says. "Want to make an appointment for tomorrow?" he asks when you are back in the office. You bet you do.

And at home, later, your family asks you, more out of politeness than real interest, "How did it go today, dear?" Now it is your turn for exaggerated cool, and you say, "It went fine. In fact, I soloed." "YOU SOLOED!" You've really got their attention now. Yes, you've now flown solo. And I don't care if you go on to become a grizzled old airline captain with tens of thousands of hours, nothing in your subsequent career as a pilot will ever quite match the magic of that first solo.

So welcome to the flier's world. Your life has become richer by a new dimension—*up*. You will find the sky a fragile ocean of oft-changing beauty. Some days it is filled with a strange landscape of cumulus clouds, mountains of white vapor around which you can playfully fly as though they had substance. Sometimes it seems as still and empty as universal space, and you can fly without a tremor of turbulence disturbing your plane. Sunset and dawn have greater beauty when you are up there in them, and few experiences can match flying high on a clear black night, with the stars surrounding you and the glowing pinpricks of lights below you.

There will be days of heat or wind, when sudden turbulence shakes you into such discomfort that flying may lose its pleasure. Or when a drab gray ceiling of stratus hangs over the world, keeping you (and your spirits) low, as the rain spits as noisily as shrapnel on the windshield. There will be days of

fog or storm when you may not fly at all.

In fine weather, when you can fly relaxed, an airplane gives you a god's-eye, mountaintop view of the earth below, and a viewpoint you can shift at your whim. You can see clearly how rivers cut their way through a line of hills by seeking always the line of least resistance; how cities sprang up where ancient travel routes crossed, or around large, safe anchorages on the coast. You can cross forests, estuaries, deserts, or tumbling mountain ranges as though they were no barrier, for they no longer are. You can see a vast steel bridge rendered (by your altitude) insignificant as a thread, skyscrapers turned to neat geometric anthills, your own substantial home reduced to less than a toy. You can travel in the straightest lines, at two miles a minute.

You can drive out to the airport with your mind beset by mundane worries and intractable problems and find that, once airborne, you have truly left them behind. You will take unimagined pleasure in mastering new skills as you learn more about flying. (I remember a friend who returned from the airport one day nearly overflowing with joy. When I asked him why, he said, "Today I flew really *well*." Another time he came back glaring furiously. "I flew *atrociously*" was all he would say. One of the things flying teaches is a perceptive new candor in self-appraisal.)

You never really do "learn" to fly. It is not a process with a defined completion. So long as you fly, no matter how many hours pile up in your log, you continue to learn. Even the private pilot's license you covet so much when you begin is itself only a beginning, and the old hand who finally issues it to you may say as he does so, "Now you can really start learning how to fly." You'll remember those words more than once—when you are wrestling your way down the approach to land at some strange and unfamiliar airport in a vicious cross wind, perhaps, or the first time you blunder into deteriorating weather and suddenly realize that you are out of your depth. (Weather he is unable to cope with is the most common hazard the new pilot—and the not-so-new pilot—faces.) Flying is safe enough, given reasonable intelligence and concentration, but it can be unforgiving of conceit or carelessness.

21

2

nk D VII 286/18

Beginnings

The urge to fly is as old as man—maybe older. Have you ever watched a cat stalking a bird? When the bird finally flutters up out of harm's way, the cat's envy of this marvelous ability is obvious. Even our earliest legends are full of magical, superhuman beings with wings (such as the cherubim and seraphim of the Old Testament) and of ecstatic ascents into heaven (which was always *in the heavens*, the sky). In 1648 an English bishop, John Wilkins of Chester, wrote that there were four ways "whereby this flying in the air hath been or may be attempted," and this approach, "by Spirits or Angels," he put as number one. Well, if you were a bishop, divine intervention was no doubt a very real possibility. His second method was "by the help of Fowlls," and don't think this hasn't been tried, with the would-be aviator seated in a light chariot to which were tethered numbers of geese, hawks, or even eagles—any birds reckoned to be strong fliers. King Kai Kawus of Persia is supposed to have tried it as long ago as 1500 B.C. If you believe a tenth-century Persian manuscript (and I don't), he did achieve a lift-off, but in the end "the whole fabric came tumbling down from the sky, and fell upon a dreary solitude in the Kingdom of Chin" —making it one of the first aircraft accidents in legend.

Icarus was the victim of another. He was the son of Daedalus, who built the Cretan labyrinth for King Minos, only to be imprisoned for his pains. Daedalus constructed wings of feathers and wax, but Icarus ignored his father's detailed instructions (as sons will) and flew too near the sun. The wax fastenings melted, and Icarus crashed to his death in the sea. We know that the air actually gets colder the higher you go, but it's a nice story, full of weird morals. This technique—"wings fastened immediately to the body"—was Bishop Wilkins' third suggestion. It has proved very popular through the centuries, and has recently been revived in the astonishing sport of hang-gliding. Dangerous, though: the fatalities have been numerous, even among modern bird-men. Ancient exponents tended to be megalomaniacs, men who just woke up one morning *knowing* they could fly, almost by faith alone. They would construct a flimsy set of sails and, without even testing their apparatus, launch forth from the highest available point. One such was King Bladud,

a legendary monarch of Britain, founder of the city of Bath and father of Shakespeare's mad King Lear:

By Neckromanticke Artes, to fly he sought:
As from a Towre he sought to scale the Sky,
He brake his necke, because he soared too high.

Or, more likely, he just fell. A luckier tower jumper was the Abbot of Tungland, John Damian, who in 1507 jumped from the walls of Stirling Castle in Scotland. He simply "fell to the ground and brak his thee bane." If he was lucky enough to survive, compound leg fractures were about the best a tower jumper could expect—unless he took the precaution of jumping over water, with an air-sea rescue boatman standing by. One who did was the town tailor of Ulm, Albrecht Berblinger, who in 1811 made his jump over the Danube, from which he was ignominiously fished, unharmed but (if an engraving of the event is accurate) making a terrible face.

Bishop Wilkins' fourth suggestion was "by a flying Chariot." Vague, but, in a way, how it was eventually done. Man first flew with balloons in the 1780s. Hydrogen gas had been discovered by Henry Cavendish in 1766; soon after, two other experimenters showed its lifting power by filling soap bubbles with it. But in fact it was hot-air balloons that flew first. They were constructed by two French papermakers, the brothers Joseph and Étienne Montgolfier, who (at least at first) had no idea why hot air rose. After early experiments with models, they began working with bigger balloons, first with animals in the car, then, keeping the balloon tethered to the ground, with men aboard. Finally after several years of this nonsense, the first manned free flight was made on November 21, 1783, the balloon rising from the Bois de Boulogne and sailing high across the roofs of Paris. The crew were a doctor named Pilâtre de Rozier and a nobleman, the Marquis d'Arlandes. A bare ten days later the first manned hydrogen balloon flew, also from Paris. The hydrogen experimenters were always rather more scientific than the hot-air makers. It is recorded that the superstitious Montgolfiers would only use chopped straw and kindling for fuel, believing that these generated a special lifting gas when burned!

At that time, ballooning seemed to be the only way to fly. Yet many of the elements the Wright

Blériot Type XI flew English
Channel in 1909. Left:
Its 25-hp Anzani engine. Right:
Hand-carved, laminated
Chauvière prop was best of
its time. Below: Blériots are
still flying in 1970s.
Lacking ailerons, they depend
on wing-warping for lateral
control. Landing gear
swivels like casters to
hold straight path on
runway in cross wind.

Top: Type XI aloft. It is so slow
that extra drag from exposed frame doesn't
matter. Bottom: Blériot departs French coast.
He left Calais on a misty day without a
compass, was lucky to find England.
XIs were observation planes in war.

brothers eventually used in their first controlled powered flight in 1903 had been around for a long time. The windmill was invented in about 1290; there are illustrations of draw-string helicopter toys going back to 1320; and kites came to Europe from China in about 1430. Of course, *powered* flight had to await the development of light and efficient engines, but even the Wrights allowed that it could have been done half a century before they did it. Gliding might have been done a great deal earlier, too. The Rogallo wing shape most popular with hang-gliding freaks is simple enough: a tubular A-frame with one cross-brace, a fabric cover, a harness to hang by, and a simple frame bar in front by which the wing can be tilted for steering. Given some sailcloth, a few bamboo poles, some stout twine, and tacks—materials available to any Renaissance gentleman, or even to a Roman centurion—construction of a hang-glider would have been easy. In truth, many of man's greatest inventions have by no means been made at the first moment they were possible. Is it perhaps still so?

Many fine minds thought and wrote about the possibility of mechanical flight, among them Leonardo da Vinci. But his thinking was perhaps typical in that he scorned gliders and envisaged either man- or spring-powered helicopters or flapping-wing machines like huge mechanical birds. Man's musculature is insufficient to develop the power required to lift his weight. In a way, mechanical flight had to await the wider dissemination of knowledge that came with the publishing explosion of modern times. Even Leonardo's sketches and notes were not known to the world until a hundred years ago.

The Wrights' success sprang from their own patience and perseverance. They were painstaking experimenters, first voraciously reading everything written about flying, then conducting experiments with their own little wind tunnel, or with models and kites (and discovering along the way that almost everything they had read about the science of flight was wrong). Determined not to give up until they had succeeded, they devoted all their spare time and summer vacations to aviation. Did their predecessors get bored more quickly, and move on to other tasks? I think so.

News of the Wrights' successes wasn't generally known until 1908, when Wilbur sailed for France with their latest "Flyer" and flew it almost daily in good weather, openly, in public, for all to see. Others had flown by then, or at least made short, largely uncontrolled hops, but it was Wilbur's mastery over his machine that astonished the French as they watched him sweep around in huge, low, superbly controlled circles. These demonstrations marked the beginning of an aviation explosion in France.

It was one of the French experimenters, Louis Blériot, who worked out the basic configuration for small aircraft that endures to this day: the pilot seated in a chair, working the left-right hinged rudder by foot pedals, and holding a stick, hinged at its base, that is tilted fore and aft for up-down control in pitch, and left-right to operate differentially movable surfaces on the wings for control in roll. (The Wrights' controls were far more complex and less logical.) It was also Blériot who put the tail surfaces at the back and who decided to mount the engine at the front of a streamlined "fuselage" with the tractor propeller attached directly to the crankshaft. This was far less elegant than the Wrights' wing-mounted engine driving counterrotating pusher propellers through a transmission chain, but it was hugely simpler and more practical.

Blériot made his epic crossing of the English Channel in his new type of airplane in 1909, and events moved so quickly that these machines, Type XIs, were still being used by several armies when World War I erupted, five years later. Although the military had not, for the most part, taken kindly to the airplane—all it was deemed fit for was reconnaissance—aircraft had been used in action several times before the outbreak of the Great War. The Italians flew several in the 1912 campaign in which they took Tripoli from the Ottoman Empire and made it a colony, Libya. Airplanes had also flown in the Balkan War of 1912–1913 and in the French campaign in Morocco and Algeria.

On the western front the opposing observation machines at first flew unarmed, but this didn't last long. Soon they were heaving half bricks at each other or firing with hand-held pistols and rifles.

Ring-mounted machine guns followed very quickly. Indeed, even a single enemy observation machine was enormously dangerous, for its observer could, if he was lucky, wreak enormous havoc on your forces with good directions to his guns.

Aerial bombing began quickly—very quickly, for on the afternoon of August 30, 1914, a little German airplane dropped three bombs on the heart of Paris, killing two people and injuring others. This unpleasant visitation became a daily event, and was called by the Parisians the "six o'clock *Taube*" after the German aircraft of that name, Taube, or dove, as inappropriate a designation as any airplane ever had.

The multiengined bomber aircraft proper was the invention of the Russian genius Igor Sikorsky, whose first four-engined machine, Le Grand (French was the fashionable language among the Russian elite), had flown before the war began. It had even been destroyed before the war began, in one of the more bizarre accidents in aviation history. The engine of another airplane flying overhead vibrated off its mounts and fell a thousand feet onto Le Grand as it sat parked on an airfield.

Sikorsky developed an improved bomber, the Ilya Mourametz. Seventy-five were built during the war, and they flew some four hundred missions. The Germans also developed huge bombers—really huge, several being larger in span than World War II bombers. The German government specified that these giants had to be well-equipped and large enough for their crews to move about inside in order to defend and service them in flight. This edict led to some monstrous designs, with the engines "buried" inside the structure and driving the propellers through complicated transmission systems that often broke down.

These huge, very slow biplanes were scarcely cost-effective weapons. Consider, for example, the Staaken R-VI: 138 feet (42 m) in span (thirty-four feet wider than a World War II B-17), with four heavy, 245-hp, water-cooled engines driving fourteen-foot propellers; a gross weight of thirteen tons, of which two tons were useful load, including maybe one ton of bombs; and a maximum speed of only 80 mph (129 km/h), giving a cruising speed of perhaps 65 mph. These bombers carried good defensive armament, with two Parabellums mounted on the front gun position as well as ventral and dorsal gun positions aft. But they were still so slow and ponderous as to be easy meat for Allied fighters or "Archie" (antiaircraft fire). They flew often at night, but they were not safe even then, for once caught in a searchlight beam they were in trouble. And since they navigated visually and had no means of making an instrument letdown, many pilots crashed simply because they were unable to find their way home after a raid. Many more were victims of accidents traceable to the unreliability of the bombers' engines.

The best of these giant German bombers were built by the Zeppelin Werke Staaken, formed by the same Ferdinand von Zeppelin who built the great dirigibles. These too were used as bombers. They had far greater range and endurance than airplanes, but proved disastrously easy to set on fire. The Germans launched both airplane and airship raids against Britain, which greatly agitated the populace but did little damage. In an offensive lasting about a year, they flew twenty-seven raids across the Channel with twin-engined Gotha aircraft, mostly at night, dropping a total of a hundred tons of bombs, killing 835 Britons, doing about $7.5 million damage. Sixty aircraft were lost, twenty-four shot down, and thirty-four destroyed in accidents.

The pure fighter aircraft (then called "scouts") followed quickly on the development of interrupter gears, which allowed a single-seater to carry a fixed machine gun that could be aimed by turning the whole airplane and firing between the whirling propeller blades—a novel concept when the Dutch engineer Anthony Fokker first demonstrated it to the Germans. These scouts were used primarily for attacking enemy observation machines, but aerial battles with enemy scouts were inevitable. The highest-scoring scout pilot of the four-year war was Baron Manfred von Richthofen, who had eighty "kills" before he in turn was shot down and killed.

Though the World War I fighter pilot has been romanticized by many, there was little real glamour or gallantry in what he did. Generally his mission consisted of making a long dive upon his prey, closing to point-blank range below and behind, and killing his victim before he himself was ever seen. He flew in open cockpits, as high as 20,000 feet (6,096 m)—without oxygen—he had no parachute or real armor, and if he was shot down he very likely died in flames. The life expectancy of new pilots joining a squadron at the war's height was a matter of weeks. It was perhaps better than dying like a rat in the mud of the trenches, but not much.

Nor had aircraft design yet become notably scientific. Wrote Tommy Sopwith, designer of the best British scouts: "Development was so fast! We literally thought of and designed and flew the airplanes in a space of about six or eight weeks. Now it takes approximately the same number of years. From sketches the designs went to chalk on the wall. Until about the middle of the war there was no stressing at all. Everything was built entirely by eye.

That's why there were so many structural failures. We didn't start to stress airplanes at all seriously until the Camel, in 1917."

The United States entered the war late, and without combat planes of her own. The United States Air Service flew French, British, and Italian designs. A former racing driver from Columbus, Ohio, Eddie Rickenbacker, was the highest-scoring American ace. His total of twenty-six German aircraft shot down is remarkable, considering the short span of time in which it was achieved. Another outstanding American airman was General William L. "Billy" Mitchell, who at the war's end commanded ninety squadrons, half of them French. Mitchell pioneered ground attack by his aircraft behind the German lines, and even planned a parachute drop of an entire army division. The war ended before this plan

D-VII replica built in France for film work.
D-VIIs were easy to fly and maintained performance
at high altitude. Top speed was 120 mph (193 km).
Color-coding of German planes was begun by the
Red Baron, von Richthofen. Designs were
heraldic, rather than for camouflage.
Bottom: Young designer Anthony Fokker (c).
Lt. Hermann Goering (r) scored 22 kills.

could be tried, and also before the vast might of American industrial mass production could make itself felt.

Before the war, aviation had been the preserve of a handful of generally wealthy experimenters. When peace returned there were thousands of young men with flying experience. Those badly bitten by the bug scratched a living in the twenties by barnstorming—wandering the land with an old Jenny or Standard military-training airplane, selling rides or stunting. One such was Charles Lindbergh, whose dream of recognition for a fantastic flying feat became possible with the development of the Wright Whirlwind, the first truly reliable aero engine. And fantastic his feat was: flying nonstop from New York to Paris in thirty-three hours in a single-engined plane built to his requirements especially for the flight. "Slim" Lindbergh's epic journey fascinated the public and inspired general awareness of what aviation might eventually achieve. The Great Depression stunted its growth, but not interest or confidence in its eventual success.

Licenses for pilots and aircraft were introduced at the end of the twenties, as aviation moved out of the barnstorming era and began to gain respectability. Developments in gyro flight attitude instruments and radio sets for navigation made scheduled air-transport services possible. (The earlier U.S. air mail service had shown the dangerous limitations of purely visual flight. Flying open-cockpit biplanes, without modern radio aids or flight instruments, the air mail service lost thirty-one of its first forty pilots in crashes.) Air racing, around pylons, coast-to-coast, or even halfway around the earth, thrilled the public. Airplanes designed specifically to suit the needs of owner-pilots began to be manufactured.

Throughout the twenties and thirties, even without the impetus of war, technical aspects of aviation advanced rapidly. The movable wing flap, allowing slower landing and takeoff speeds, was invented by Hugo Junkers in Germany in 1919. His company also pioneered the metal-skinned airplane. The first practical retracting landing gear appeared in 1920 on the Dayton-Wright RB. Soaring in gliders began in Germany in the early twenties, and the

first motorless flight of more than an hour took place in 1922. The Spanish designer Juan de la Cierva flew his first Autogiro—forerunner of the helicopter—in 1923. The variable-pitch, constant-speed propeller appeared in 1924. The first proposals for jet engines were published in 1926, even as the first liquid-fueled rocket flew in America. In 1926, an American described how aircraft skins could be "stressed" to become an integral part of the structure, rather than just a covering for the load-bearing skeleton. The first modern monoplane airliners with real performance appeared in 1933. The seemingly immortal DC-3 entered service in 1936, and the first pressurized transport, the Boeing 307, flew in 1938. The first remotely practical helicopter, the German Focke-Achgelis FW-61, flew in 1936, and the first practical single-rotor helicopter, the work of the same Igor Sikorsky who pioneered the big bomber in 1913, flew in the United States in 1940. The first jet airplane and (liquid-fueled) rocket airplanes flew in 1939.

The story of Sikorsky's early helicopter work is an appealing one. The idea had been in his mind since 1909, when he first tried to build a prototype in czarist Russia. Some thirty years later the United Aircraft Corporation, his parent company in the United States, decided to end production of the elaborate, baroque flying boats he had been designing and building for Pan American Airways. Officials tried to break the news to him as gently as possible. He seemed miles away as they spoke; they feared he might burst into tears. Finally he asked, in his still-thick Russian accent, if he might be allowed instead to try a "heelicopeeter." Relieved that his whim was so modest, and wishing to humor him, they said sure. In 1939 he pushed out into the sunshine his Vought Sikorsky VS-300—the "Ugly Duckling," he called it—a strange-looking tangle of welded steel tubes, with a bucket seat at the front, the main rotor attached to a stalk above the bare engine, and a tiny tail rotor at the end of a bridgework truss behind. Still wearing his neat gray business suit, a silver tie neatly clipped to his white shirt, and a felt hat with the brim jauntily turned up at the front, Sikorsky climbed aboard to test-fly it. Later, in demonstrations for visiting dignitaries, he hovered with amaz-

Preceding pages: Three rare air-to-air
photos from World War I. German fighter (top)
is Fokker Eindecker (monoplane) which
dominated air combat in late 1915.
Biplanes are Sopwith Camels, a stocky,
aerobatic type that downed 1,294 enemy craft.
Right: Close-up of Eindecker, first to have
synchronized machine gun firing through prop.

ing precision, then danced ponderously backward, forward, sideways. He speared the nose of his craft into a circus ring held by an aid; he hovered with a dozen eggs suspended in a net from the landing gear, lowered them gently to the ground, and then landed in a small, fenced enclosure, carefully shattering one of the eggs to prove they weren't hardboiled. This last whim, this Ugly Duckling, provided the basis for the flood of helicopter technology that has happened since.

Aviation between the wars is remembered most vividly for record flights, spanning oceans, girdling the earth, racing the clock, and all the other newspaper clichés. If the public wanted heroes from aviation, the newspapers supplied them. But for every "Lucky" Lindbergh, there were scores of other pilots, forgotten now, who sought glory and found only oblivion, too often literally, as they vanished into the oceans, leaving no trace of what had befallen them. Losing control while flying in cloud and falling into a spin was perhaps the commonest cause. Lindbergh himself told how he fell into a spin halfway across the Atlantic, recovering barely above the waves. To tired pilots of lesser ability and experience, flying by the primitive instruments of the time, it was a perilously easy trap to fall into.

Lindbergh's transatlantic flight was minutely planned. Many who tried to follow him were desperately ignorant and unprepared. Two men who made it, but by quite different approaches, were Howard Hughes and Douglas "Wrong Way" Corrigan. Hughes, the strange, antisocial multimillionaire, dabbled in movies, movie stars, and speed in the air. In 1936 he held the world airspeed record at 352.338 mph (567 km/h) in a much-modified mail plane with a 14-cylinder, 1,000-hp Wasp engine. The craft had cost him some $120,000 to build. In 1937 he flew it, solo, from Los Angeles to Newark in seven and a half hours, breaking every record for the trip, including those he himself had set the previous year.

In 1938 he decided on one last, dazzling, long-distance flight, the ultimate one: around the world. He bought a twin-engined Lockheed Super Electra, equipped it with the most advanced gyro instruments, autopilot, navigational devices, and radios, and hired a superbly skilled four-man crew to fly

with him, working in shifts. He agreed to help promote the forthcoming New York World's Fair through the flight in exchange for permission to build, on the fairgrounds, a radio tower powerful enough to relay to him weather reports based on a thousand daily observations around the world. He took off from New York's Floyd Bennett Field one summer evening and was back in less than four days, having flown nearly fifteen thousand miles (24,135 km), never more than six miles (9.6 km) off course, and with the flight going as smoothly as clockwork. He had cut in half the old around-the-world record, set by Wiley Post in 1933.

Corrigan, unlike Hughes, didn't come from money. In fact, he was something of a drifter, never staying at one job for long. He had been a mechanic at the Ryan plant, working on Lindbergh's *Spirit of St. Louis*, when the great man took time to shake his hand and wish him well. "That handshake was my inspiration to keep trying," he said, but at exactly what he was too cagey to say. What he was planning was nothing less than to duplicate Lindbergh's famous flight. In 1933 he bought a secondhand Curtiss Robin for $325. During the next three years he acquired a used Whirlwind engine for it, refabricked it, and installed a huge gas tank in the cabin, obscuring his forward view.

By this time there had been so many tragedies among amateur Lindberghs that the United States government would no longer permit transatlantic jaunts. But Corrigan was determined to get around the ban, and in 1937 a wild idea occurred to him: he would land at Floyd Bennett Field after officials had quit for the day, gas up quietly, and go on without permission. The following summer he sought, as cover, a permit for a nonstop flight from Los Angeles to New York and back again. He was grudgingly granted the first half; the permit for the return trip would be issued only if he actually made New York nonstop on the first leg. He arrived with four gallons of gas left in his tanks, and then took a week's vacation. At first nobody even realized he had arrived nonstop from California in a nine-year-old crate with minimum instruments and no radio. The newspapers were full of Howard Hughes's world-record flight, and when they did take notice of

Corrigan's transcontinental jaunt it was only to draw comparisons between his ancient Robin and Hughes's superbly prepared Lockheed.

Nine days after his arrival in New York, Corrigan tanked up and took off at dawn, still supposedly for California. A long time later, having seen little but fog and gasoline sloshing about his shoes from a leak in his tank, he touched down at an airfield near Dublin. Forever after he maintained his spoof, insisting that his compass had misled him, that the endless clouds had prevented him from identifying any landmarks, that he had mistaken Boston for Baltimore. The world (except the U.S. Department of Commerce, whose regulations he had so manifestly disregarded) laughed and laughed. The Department admired his "daring and skill, and the world-wide sentiments of good will" he had created, but all the same they suspended his pilot's license until he had returned to the United States by ship, just in case he had any future plans.

Corrigan, whose take-home pay as a mechanic had been $50 a month, cleared $75,000 for his escapade, what with a book (which he wrote himself, after firing the ghost writer assigned) and sale of the movie rights to it. He used the money to buy a fruit farm in California and retired forever from the limelight. For him, as for Lindbergh, the gamble had been for high stakes, but it had succeeded.

The Battle of Britain in 1940 was also for high stakes, in fact the highest—survival of a free Britain. And if the German blitzkrieg in Poland had shown how ruthless bombing could shatter a country's defenses, the RAF in 1940 showed that resolute fighter defenses could halt a conquest. It was a small battle. No more than two thousand pilots fought on the English side, and fewer than five hundred died. And so spread out were the defenders, between their fighter bases, radars, and controlling headquarters, that the titanic importance of the battle was lost on them individually. They just flew and fought by day, and retired to the mess to drink beer at night. Their Spitfires and Hurricanes and the German Messerschmitts were small single-seaters, armed with machine guns or small cannon, and powered by slim, liquid-cooled, V-8 engines of about 1,000 hp. They had a maximum speed of about 350 mph (563 km/h)

and fought mostly at medium altitudes, from 20,000 feet (6,096 m) down. They were unpressurized, their pilots breathing oxygen through face masks. The primary role of the defending British fighters was to attack the formations of German bombers, which forced the Luftwaffe to send larger and larger forces of fighters to accompany and defend them. Without radar the RAF could not have won. It enabled them to estimate the size and probable targets of incoming raids in time to concentrate fighters in their path. The alternative to radar would have been standing patrols, requiring impossible numbers of aircraft and trained pilots.

By the end of 1940 the RAF Fighter Command's "kill" rate on German bombers forced the Luftwaffe to switch to night bombing. For a long time the RAF was powerless to stop night raids. Early radar was heavy and cumbersome, and the development of sufficiently miniaturized airborne sets to fit into even twin-engined fighters took time. That the day bomber could not survive against fighter opposition was something the RAF had already learned in the Battle of France and in an early attempt to bomb Berlin.

By 1941 the prospects for victory looked remote for the British Commonwealth countries still fighting Germany, Japan, and Italy. Then Germany turned on the Soviet Union, and Japan launched a devastating surprise attack on the American base at Pearl Harbor in the Hawaiian Islands. Thus the two greatest powers, hitherto neutral, were brought into the war on the Allied side. Germany and Japan now had ranged against them a might of manpower and manufacturing industry that would devastate them within four years.

The United States, far removed during the 1920s and 1930s from the growing troubles in Asia and Europe, and beset by economic problems of her own, was ill-prepared for war—as Britain had been two years earlier. American warplanes generally were no match for the enemy. U.S. Navy aircraft were shot from the Pacific skies by the nimble Japanese Zeros, and early U.S. fighters supplied to the RAF under lend-lease could not hold their own against the Messerschmitt 109s. Even the early versions of the famous Boeing B-17 Flying Fortress

41

Left: Sopwith Pup of 1916 chases Fokker DR-I Triplane. Below: Cockpit of Triplane replica has few instruments. LVG C-VI, first German recon plane and light bomber, made bold daylight raid on London in 1916, dropping six tiny bombs. Plans for Pup still can be bought from Hawker Siddeley.

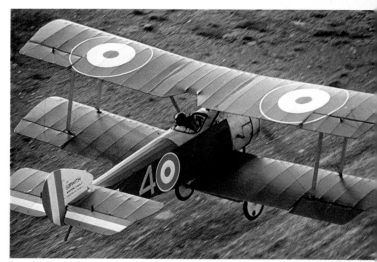

Two Firsts: Ryan's *Spirit of St. Louis*
was designed and built for Charles
Lindbergh's epochal nonstop transatlantic
flight in 1927. A flying fuel tank
(with no direct forward visibility), it had an
amazing 4,210-mile (6,774-km) range.
Bottom: Igor Sikorsky demonstrates his
prototype VS-300 helicopter in 1940.

were hardly fit for combat. They burned too easily, suffered too many systems failures, were under-armed with .30-caliber guns, and were cold and drafty. Boeing quickly improved the B-17, equipping it with .50-caliber guns, power-operated turrets, and better armor. Even so, when the Eighth Air Force began bombing Germany from British bases in 1943 it found—as had the European fighting powers earlier—that daylight bombing against well-defended targets could not be sustained. In 1943 the Luftwaffe could launch a thousand defending fighters at once, and U.S. bomber losses rose as much as 120 percent (basic complement plus replacements) within four months. The bombers were switched to less well-defended targets in France until enough fighters with the range to accompany the Fortress fleets to Germany and back could be built. These were Thunderbolts and Mustangs—the latter perhaps the best fighter of the war, with a 1,000-mile (1,609 km) range and superb performance at altitude from its highly supercharged engine.

Although the B-17 Fortress is the best-remembered U.S. bomber of the war, it was not the most numerous. Half again as many Consolidated B-24 Liberators were built. These American formations, bombing by day, and matched by RAF Halifaxes and Lancasters flying by night, wrought increasing devastation upon Germany. Particularly effective were the attacks on German oil installations and rail transportation. And so effective were raids on German fighter factories that by war's end German fighter defenses were almost nonexistent.

If World War I had been a technological struggle for supremacy, World War II was still more so. In the quest for ever more performance fighters, both Britain and Germany developed jet engines. The British jets employed centrifugal impellers, a technological blind alley that was abandoned soon after the war's end—and, indeed, British jets were hardly in service in time to have much effect. The Germans pursued the axial-flow compressor, such as modern jet engines have, but because of political pressures and the relentless Allied bombing, they never were able to make their engines reliable enough for their few jet fighters to be truly effective.

Both sides were energetic in developing electronic devices. The British innovations in early-warning radar were soon copied by the Germans, and both sides went on to evolve radio beams to guide night bombers as well as radar for ground mapping, bomb aiming, and airborne interception by night fighters. These necessitated a degree of miniaturization that had looked impossible at the war's start and which, considering the total reliance on bulky thermionic valves in those pretransistor days, was a formidable achievement. Each side was hoping to achieve some technological breakthrough so cosmic in its nature that it alone would end the war in the inventing nation's favor. In the end, only the United States was able to do this, with the atomic bomb that resulted from the hugely expensive Manhattan Project, though by that time the conclusion of the war was no longer in doubt.

The Japanese, however, did not pursue technical advances with such vigor. Their Zeros proved to be vulnerable to attack owing to their lack of armor plate and fuel-system protection. Once the U.S. Navy had better fighters—the Chance Vought Corsair and the Grumman F4F and F6F, for example, strong, powerfully engined and well-armored aircraft—the Americans began to drive the Japanese forces back toward their mainland, island by island, battle by battle.

The ultimate bomber of the war was the Boeing B-29 Superfortress, twice as powerful and heavy as the B-17, with a pressurized cockpit to enable crews to work in comfort. Every gun was power-operated by remote control from the crew compartment. The B-29's 300-mph (483 km/h) cruising speed was almost twice that of the B-17. The 29s were not ready in time to be used in Europe, but they were able to burn the heart out of Japan's cities and industries. Japan's fighter and gun defenses were not nearly so strong as Germany's, and the relatively few B-29 casualties resulted from running out of fuel toward the end of a raid that usually involved a 3,000-mile (4,827 km) round trip over water, or from engine fires, for the airplane had been rushed into service without enough time for prolonged testing.

Only the Germans made big strides with their pilotless missiles and rocket propulsion. Their R4M air-to-air missiles were highly effective, and their

Workhorses of World War II: B-17 Flying Fortress
(above & below) was Allies' principal daylight bomber
in Europe. Britain's Spitfire (r & below) was a
superb airplane, went through 24 models in
course of war. Marks I & II fought Battle of Britain.
Underview (r) shows unique elliptical wing.
P-51 Mustang (top r) was America's best fighter in all
theaters. Grumman F4F Wildcat bore brunt of
early Pacific campaign for U.S. Navy.

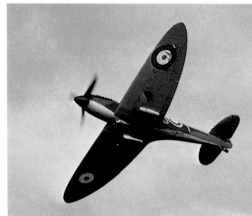

V-1 pulse-jet, medium-range missiles had an undoubted effect on British morale far above any actual damage they caused. Their A-4 or V-2 missiles carried a ton of high explosives and were the first ballistic rocket missiles.

The German Me-163 Komet was the only rocket-powered fighter of the war and the first aircraft to fly faster than 1,000 km/h (621 mph) in level flight. But time did not permit its development as an effective defense weapon. At war's end the Germans were further advanced in high-speed and transonic flight. They had built transonic wind tunnels for research and discovered the virtues of wing sweepback in delaying the adverse effects of locally supersonic airflow at airspeeds approaching that of sound.

But in the end the war was decided by manufacturing capability and population size. The Allies were able to build more weapons and train more combat forces than the Axis nations. Had all else failed, the decision would have been made by the nuclear bomb, which, in a curious way, was one outcome of Hitler's political beliefs. The atomic bomb was the brainchild largely of European refugees from Nazism, living and working in the United States.

America's late entry into the war and remoteness from the fields of battle gave her one enormous advantage in postwar years. She had been able to produce the finest transport aircraft in the world during the war and she was able to supply these to the world's airlines when peace returned. This established a commercial lead which endures to this day. It began with the DC series in the 1930s, twin-engined airliners with light but powerful and reliable radial engines, strong, well-engineered airframes, and a cruise performance that gave them real advantages over the railroads. The DC-3 remains the transport aircraft that was built in the largest numbers (eighteen thousand, including military versions), and although its design is more than half as old as the entire history of aviation, there are probably still more DC-3s still flying than any other airliner type.

Air transportation got a chance to show its capabilities very soon after the war, when the Russians cut the roads giving access to the Allied sec-

tors of Berlin from the west. The Allies organized a Berlin airlift to supply the city's total needs by air, the first time a city had been victualed in this way. The airlift was expensive, but it worked, and West Berlin is still western in consequence.

Both the United States and the USSR abducted German scientists who had worked in rocketry and high-speed flight, and having picked their brains allowed them to return home. From this humble and perhaps ignominious beginning evolved supersonics, jet transports, and the space programs. The United States built the first supersonic research aircraft, the rocket-powered Bell X-1, in 1947. The Soviet Union launched the first earth satellite, a tiny, grapefruit-sized object capable of emitting only a beeping radio tone, in 1957. When a big international war erupted again in 1950, both the United States and the USSR had swept-wing jet aircraft, the North American F-86 Sabrejet and the MiG-15. The Russian fighter was lighter and could climb faster and higher, but the American had better transonic performance and weaponry, and most decisively won the air battles over Korea.

By modern standards, those early jet engines were heavy, thirsty for fuel, and low in power. But slowly they were improved. It was the British in the late 1940s who first decided jet power-plant performance was adequate for a four-jet transport. Even so, the de Havilland Comet airliner had to have a light structure to achieve satisfactory performance, and tragically its airframe proved too light to withstand the weakening of metal fatigue—an insidious effect, hard to detect, resulting from repeated reversals of stress on highly stressed structural components. Fatigue had first been experienced in the axles of railroad cars in the mid-nineteenth century, but its devastating effects on aircraft structures had not been foreseen. Fatigue ruinously shortened the useful lives of many postwar transport aircraft and is a problem still. Fortuitously, the DC-3 almost alone avoided fatigue problems through being unpressurized and having a multicellular box-spar wing structure, with no individual spar member being highly stressed.

The great American transport-airplane builders, Boeing and Douglas, were able to bide their

49

time till jet engines had improved still further, and in the 1950s to develop the 707s and DC-8s that were the first of the line of jet transports that almost all the world's airlines fly today. Fan engines, wherein a portion of the engine's compressor airflow by-passes the "hot" section and combustion chambers, and forms a cool efflux shell isolating the hot, high-speed gases from the engine core, have given jet transports hugely better range, lower fuel consumption, and a relative quietness that populations living near airports can tolerate. Present-day jumbos —the 747, DC-10, Tristar, and A300 B—were all made possible by the superior performance of a new generation of huge fan engines.

The discovery in the 1950s of the germanium transistor has enabled avionics—aircraft radio equipment for communications, navigation, and letdown, plus automatic pilots and aircraft computers—to make enormous progress. At the end of World War II, there existed only complicated loran oceanic navigation sets, primitive instrument-landing systems, and radar ground-controlled approach systems. Today fully automatic letdown and landing devices are in service. Inertial navigators originally developed for ballistic rocket missiles now steer aircraft over the longest distances with uncanny accuracy. Modern airliners are for much of the time flown by computers, with pilots monitoring the operation. A World War II device called IFF—Identification Friend or Foe—has been worked into radar transponders that identify an aircraft on ground radar and automatically transmit its altitude. And even the more

basic navigation equipment has progressed enormously from the early radio compasses, which required the pilot to turn his antenna to find a signal "null" in order to get a bearing from the ground station.

Many of these electronic devices are now small enough and cheap enough to install in private planes. The growth of private and business flying, now generically described, along with air-taxi and agricultural aviation, as "general aviation," has been a post–World War II phenomenon. By the mid-seventies, nearly a million Americans were either student pilots or fully licensed to fly, and general aviation was carrying more people within the United States than were the airlines. This growth of general aviation was foreseen by most manufacturers of small aircraft at the end of the war, but it took longer than they expected, and some of them went bankrupt producing private aircraft that the returning heroes could not afford. I think the prosperity of postwar America was responsible for the eventual boom in private flying. Ever since Lindbergh's memorable flight made people aware of aviation's possibilities, there have been more people wanting to become pilots than could afford to. The growth of general aviation has also been encouraged by the airlines' purchase of increasingly larger-capacity aircraft, which they can keep filled only by flights between the largest cities. The energy crisis has also contributed. Private planes, particularly the smaller ones, are more efficient fuel users than Detroit's ground-bound behemoths. Finally, reliable avionics

Extending Boundaries of the Flier's World:
Saturn V rocket, most powerful engine ever made,
lifts off from Kennedy Space Center at
start of Apollo 15 mission, the
seventh manned lunar voyage, in 1971.

and more sophisticated pilot training, with the best private pilots as skilled as airline pilots at instrument flight in weather, have made travel by private plane more practical.

Flying has become a marvelous sport, a recreation and hobby for modest armies of enthusiasts. One cult recaptures the delights of Montgolfier ballooning, drifting slowly downwind under colorful envelopes filled with hot air from simple propanegas burners. Another group, in love with history, seeks out rusty, hay-covered airplane skeletons in farmers' barns and painstakingly restores, renovates, re-covers, and revarnishes them till they gleam more brightly than the day they first left the factory decades ago. For some, aerobatics represents the ultimate in piloting skill; devotees spend every spare moment practicing uncomfortable gyrations for the next contest. Once a largely European sport in which the Socialist countries of eastern Europe excelled, aerobatics has been taken up by Americans with such enthusiasm that the United States finally won the world championship in 1972.

Many amateur pilots like to construct and even design their own aircraft, often using their home cellars or garages as workshops. The designs they favor reflect the dreamy romanticism of the typical home-builder: open-cockpit biplanes, racing craft, aerobatic specials, replicas of World War I scouts, or layouts so weird no commercial manufacturer would touch them. (Cost, it must be admitted, is also an incentive. Home-built aircraft cost far less than "store-bought" planes.)

The latest aviation cult is a throwback in time, to even before the Wright brothers. This is hang-gliding: hanging in a simple harness from homemade wings and steering, usually by weight shifting, just as Otto Lilienthal, Percy Pilcher, and Octave Chanute did in the nineteenth century. Curiously, the hang-gliding movement has grown up as an illegitimate offspring of the space program, for the simple V-shaped sail wing most hang-gliders use was originally invented by a NASA scientist as a foldable recovery vehicle for spacecraft reentering the atmosphere. What folds to go into a spacecraft happily also folds to go on an automobile roof rack. Hang-gliding has brought about a further democrati-

zation of flying by making it affordable by even the most modestly wealthy businessman or hipster—a hang-glider kit costs only a very few hundred dollars. The fact that hang-gliding is dangerous—there have been injuries, even deaths—seems only to add to its appeal for many people.

If Pearl Harbor caught America still half-asleep in December 1941, the pressures and anxieties of the long cold war with the Soviet Union have been a goad to ensure that this will never happen again. The United States now has a commanding lead in almost every aspect of aviation and space technology, a lead born of resolve to be ahead and backed by the wealth to stay there. In high-speed flight, for example, that first supersonic rocket ship was followed by others: the Douglas Skyrocket, the Bell X-2, and the astonishing North American X-15 series, which flew so fast and so high that some of their pilots qualified—quite fairly—as the first astronauts. Much of the later technology of space flight was pioneered with the X-15 program: rocket engines, thrusters for altitude control, telemetry to ground stations, pressure suits and pressure breathing for the pilots. Moral considerations about the use of spy planes aside, one of the United States' more formidable developments was the Lockheed U-2. This aircraft flew so high—well above 60,000 feet (18,288 m)—that it was years before the Soviet Union even had a missile capable of preventing U-2s from overflying its territory at will. American airliners since 1945 have always been—even if only marginally—superior in operating costs and profitability; and profitability is the crucial factor in such a high-cost and competitive business. Though the simple performance of modern military aircraft may not vary significantly from nation to nation, the advances in their weaponry and systems have all been evolved in the United States.

But the most fantastic achievement of all has been Project Apollo, which successfully landed men on the moon and then brought them safely home—surely the single most astonishing feat of mankind. And space flight is no more than an extension of the sciences of aviation. Astronauts and cosmonauts usually are former airplane test pilots; and the instrumentation and control systems of space vehicles are offshoots of aviation usage, as are their power plants, computers, and life-support and communications systems.

Virtually the only facets of modern aviation technology the United States has so far disdained to pursue are jet vertical takeoff and commercial supersonic transports. America's early SST program became in the end a political rather than a technological issue. Almost anything in the realm of aviation is now possible, but often a new project is so expensive that the political question of whether to allocate the necessary funds determines its fate.

From that first hesitant, pitching flight by the Wrights' rickety biplane, 120 feet (37 m) across the Carolina sands, to the vast affair that aviation is today, has taken only one human life span. There are plenty of people still vigorous who were born before the Wrights flew. If man was late in discovering how to fly, he has certainly made up for it. "I am well convinced," wrote the English experimenter Sir George Cayley in 1804, "that Aerial Navigation will form a most prominent feature in the progress of civilization." And, stretching his imagination to its limits, he prophesied that we might one day fly "with a velocity of from 20 to 100 miles per hour." Today we can achieve earth-escape velocities of 18,000 mph (28,962 km/h), and carry ordinary fare-paying passengers within the atmosphere at 1,500 mph (2,414 km/h). Who would like to prophesy what aviation may have achieved in another hundred and fifty years, or even in seventy-five? Will orbital rocket airliners take us to any other part of the globe within ninety minutes? Will we all have vertical-takeoff cars, small enough to park in the street, that can navigate themselves and avoid collisions automatically? Will children go soaring on winged bicycles? It seems that aerospace technology can now achieve almost anything, subject only to the limitations of cost and our reliance on dwindling supplies of fossil fuels. (One manufacturer is already experimenting with liquid hydrogen as a fuel, thinking this may easily be prepared by the electrolysis of common water by nuclear power, or perhaps by some source of power yet to be discovered.)

Only one thing is certain about the future: it is coming.

3

Cessna Aerobat, small monoplane,
executes exuberant roll. "Most piloting is
a matter of small, delicate control inputs, allied
with a good deal of thinking about what to
do next." It is an excellent pastime
for perfectionists.

gyro instruments, feeding corrections automatically into the aileron controls. Large aircraft all have auto-pilots that can fly the machine for you. In the newer airliners they can even make automatic landings. Some modern military aircraft rely so completely on autostabilization that they would be quite unflyable without it. Even in a small airplane, an autopilot is a joy on a cross-country jaunt, leaving you free to watch for other traffic, study your charts, change radio frequencies, and so on, without having to make constant control corrections to stay on course and at a set altitude.

Check lists are a feature of aviation that seem to fascinate first-time passengers in small airplanes. "Use your check list," your instructor used to say several times an hour. And many airports have a little sign near the engine run-up point at the start of the runway that repeats the same message. But when passengers see you with that little manual open on your knee, as you run through the listed vital pre-takeoff actions one by one, they think, *Doesn't he know how to operate this plane? Why does he have to keep referring to the instruction book all the time?* Airplane salesmen know all about this. Have you ever seen a salesman on a demo ride using a check list? *Never.* It's no problem either for pilots of large aircraft, who have made a litany out of checks, one reading them out, the other answering, "Check," and who, in any case, are separated from their passengers by a closed door. If you own the plane, you quickly memorize the checks, but if you borrow or rent there's nothing for it but to work through the printed list each time for safety's sake. But your passengers will surely want to know what in hell you are doing, so it's best to explain.

A surprising proportion of passengers are profoundly uneasy about flying at all. I once took two apprehensive friends on an aerial tour of France in a small Cessna, and whenever we flew through a small bump they gasped and shrieked. When I put the plane in a moderate 45-degree bank to show them the yachts in St. Tropez harbor, I was belted on the ear with a handbag and told to return to level flight *immediately.* It can be hard to explain to such aerophobes exactly what we see in flying. There is the god's-eye view of the world it gives; and

that same joy in mastery of one's mount that a horseman feels. Flying is an activity for the achievement-oriented, for those who relish the difficulties, the book work of learning, the cost of flying time, the hard slog of that practice that in the end maketh perfect. And the knowledge that one can be killed in an airplane lends spice to mastering it and seeking to keep the hazards at bay forever. There are those who say rubbish to this aspect, who say that flying, properly conducted, is supremely safe. Yet, if so, why do all pilots delight in each other's war stories? Why do aviation magazines all print detailed accounts of horrendous accidents? "So that other pilots may learn from the mistakes of the unfortunate few," they claim. Nonsense! It's more the machismo of the bullfight, the public's enjoyment of the spectacle of the occasional unfortunate, unwary Christian getting caught by a particularly sharp lion. This enjoyment of aviation for the pleasure of triumphing over its inherent perils is perhaps strongest in pilots who have some timorousness in their psyche, which is perhaps most pilots, and certainly the author of this book.

Flying is safe enough—certainly, on a distance-traveled basis, safer than automobile travel—but the airplane is inherently unforgiving of foolishness. (A flying club in England has a little reminder stenciled on the instrument panel of members' craft: "All airplanes bite fools.") In addition there are two characteristics of the machine that catch the unwary: it cannot stay upright unaided and it has a minimum speed below which it quits flying. These are particularly unsettling to people accustomed to driving cars, which behave altogether differently. (I learned to fly as a kid, before anyone would let me near a car, so I had no trouble getting acclimated to a plane, but I did have to make adjustments later, when I learned to drive a car.) Without doubt, flying is a more complex skill than driving, and therein lies a great deal of its charm. There is always something that you can usefully be practicing. And so we do. The student or low-time private-license man spends hours on touch-and-goes, around and around, trying to achieve the perfect "greaser," exactly "on the numbers"—those big white numerals painted on the blacktop at the runway's start to indicate the

En Route to an Instrument Rating: Aloft
with instructor, pilot wears hood allowing him to
observe instrument panel, but nothing outside plane.
Flight pattern must be interpreted by what
gauges—not senses—tell him. Below: Instrument
gauges of plane at rest.

compass direction with which it is aligned. When this can be done nine times out of ten, the same level of success is sought in a diabolical cross wind, sideslipping through the wind so the wheels will touch down uncrabbed. A Chinese approach, if you like: Wun Wing Lo. (Aviation is full of terrible old jokes and sayings. "There are old pilots, and bold pilots, but no old bold pilots" is another of my favorites.)

Surely it is weather that presents pilots with their sharpest dilemmas. If you are not instrument-rated, then you are a VFR- (visual flight rules) only pilot and must by law be clear of cloud and well within sight of the ground. You are taught, for safety's sake, that when the clouds drop down and visibility diminishes you must make a 180-degree turn and head back out of it. But how bad is bad? In your early days as an airman you discover that more experienced pilots continue flying quite happily in weather you turn back from. So you start pushing your luck a bit; how else will you learn? And pride comes into it. It is galling if Joe can make it back from the weekend at Lakeville Junction in his plane, and you can't. It is equally galling to have passengers and to feel obliged because of deteriorating weather to set down somewhere a hundred miles from home. You then have to explain that the night must be spent in some Happy Hour Motel, which means we'll all be late for work tomorrow, even if the weather front does go through during the night. Even if they're polite about it, you know what they're thinking: *Him and his flying. Next time we'll drive.*

Here's the dilemma: to learn about flying in vile weather you've got to try it, but trying it is risky. What's the answer?

Perhaps to be cautious in your bravery. To nibble at weather, making certain that what you're taking on is only a little worse than what you've met before. Said the instructor who taught me most about flying: "Always have an out." Such as an airport just a little way back, where you know the weather is reasonable because you just went by and radioed them to check on it. Where you can go if you have to make that 180-degree turn. It also helps to know exactly where you are when the cloud ceiling lowers and rain spatters the windshield, obscuring your view. That's the time to have the sectional chart on your knee and to follow every detail —road, rail, river, township, and topography—with care and attention. It's a good time to be fat with fuel, too.

In inclement weather, it is easier to fly over terrain that is familiar. Presumably, you know just where the hills start and where that new thousand-foot television antenna is being erected. It is also wiser, if you have to take any chances, to take them when you are alone, with no passengers.

If you are lucky enough to fly a big twin, you've probably got pneumatic wing boots and alcohol sprays for the props, so ice hardly bothers you. Anything less, and you haven't, and it does. So light rime icing in cloud is forecast. How high are the cloud tops? Will you be able to blast up through to clear air? The height of the top of the cloud is something weather bureaus are usually vague about, unless another pilot has just flown the route and taken the trouble to report it. Suppose you pick up a load, what is the freezing level? Can you drop below it in safety to melt ice on your wings?

Thunderstorms are a bigger dilemma still, no matter how experienced you are. Storms can break airplanes, or so upset them that the pilot is unable to recover level flight intact. The thunder is no problem. It is merely noise. Nor is lightning as dangerous as it might seem, for an aircraft in flight is not earthed and is made largely of aluminum, a fine conductor. So lightning, if it touches a plane, usually leaves it unharmed. But the power surges of an airborne lightning strike can burn out transistors in the avionics, leaving you with no means of radio navigation; and lightning will also burn its way through semiconductor material, such as the plastic covers of radomes, or fiberglass wing tips. There is also at least one case on record of lightning igniting the fuel vapor in an aircraft's tanks.

Hail is trouble. It can batter an airplane and damage it expensively. But the real menace of thunderclouds for aviators is the turbulence associated with the vertical air currents inside them. The National Severe Storms Laboratory of the University of Oklahoma made more than a thousand penetrations

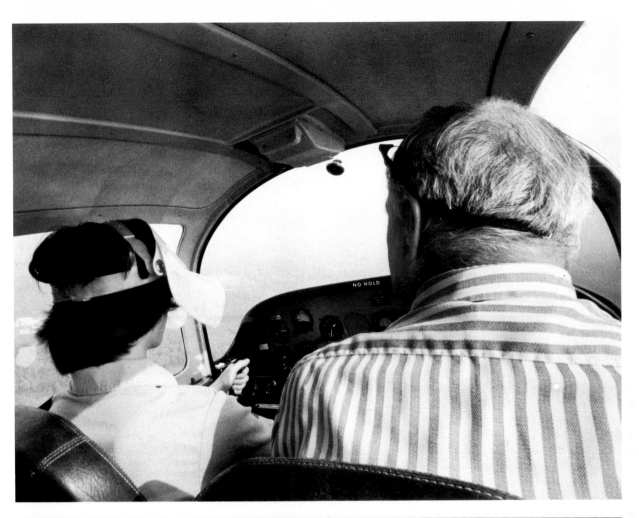

of the meanest thunderstorms it could find, using military jet-fighter aircraft. It reported that while most vertical currents were on the order of 15 or 20 feet/second (say 14 mph), they were occasionally much stronger. The maximum vertical gust measured was 278 f/s, or nearly 200 mph. The McDonnell F-4 used in the program recorded one bump of 5.5 g. These figures show a degree of turbulence sufficient to break an ordinary lightplane.

So, standard procedure for *all* pilots of *all* aircraft is to do their best to stay out of thunderstorms. Bigger aircraft (piston twins and up) have a radar set mounted in the nose to map precipitation (rain and hail), so it can be avoided. While turbulence itself does not show on radar, rain does, and if you avoid the rain and hail of a storm you also avoid the turbulence—at least, if you give the precip a wide enough berth, you do. It is now a Federal law that an airliner may not be dispatched on a flight through an area of known thunderstorm activity unless its weather radar is in good working order.

Flying light aircraft in thundery weather presents many problems. The first begins when you telephone the FSS (flight service station) for a forecast. You are likely to be given an absurdly pessimistic picture—severe storms in all quadrants, heavy ice, severe turbulence, hail, tornadoes—much of which may well be fantasy. Next you request radar reports, although these can be a couple of hours old and therefore inconclusive. In the end, experience teaches you to interpret what you are told. You learn that many FSS's seem to do their utmost to discourage you from flying if there is even the slightest threat of bad weather. You learn what sort of weather will permit you to climb above the clouds and spot thunderstorms far enough ahead to avoid them. You learn when it's best not to file IFR (Instrument Flight Rules) at all, but to stay VFR and underneath the clouds. And you learn to recognize when it's best not to try fighting your way through to your destination, but to look instead for temporary haven at some nearby airport. You may have to go out of your way, but at least you will be safe from serious turbulence.

Single-engined light aircraft and the smaller twins have no radar; the controller at Center does, if

62

Flier's map guides him through network of
airfields and beacons, away from
hazards and restricted air spaces. This detail from
a Coast & Geodetic Survey map
shows Washington-Baltimore area.

you are flying IFR and working him. It is not unreasonable to expect that he will be painting storm echoes and giving you radar vectors around them. And, indeed, he does his best and often can give you a good deal of help. But his task is to provide separation only between IFR traffic, not between the traffic and storms, and his radar uses special circular polarization circuitry to eliminate precip returns, so you cannot rely on his assistance.

As a fledgling instrument pilot I remember asking experienced aviators their views on thunderstorms. The answers I got were strikingly inconsistent. My instructor said: "You stay out of them." But a colleague who owned a Comanche, which he flew regularly between New York and Florida, said storms didn't frighten him, he'd flown through plenty. "You see some lightning flashes and you bounce a lot, but that's about all. I never met much ice, or anything that scared me." Said my instructor, when I passed this on to him: "He'll bust his ass if he makes a habit of busting thunderbumpers."

A professional Beechcraft Baron pilot, who operated all over Europe, said, "I avoid them when I can, and when I can't, well, it gets very dark, very bumpy, and rains like mad, but not for long."

Said a very senior Boeing captain: "My airplane is strong enough to take it; I never worry about that. Where radar has really helped us is in cutting down expensive hail damage—dented leading edges, cracked windshields, broken radomes, that sort of thing."

And from a British airline captain: "Before we had airborne radar, we just used to flog through them, in the old DC-6s and -7s, going down to Africa. The only really scary encounter with turbulence I ever had was in bright, clear air, crossing the Andes at 33,000 feet, and not a thunderstorm for a hundred miles."

You can provoke an all-night discussion by asking a group of airline pilots which parts of the world have the worst thunderstorms. Equatorial Africa and Latin America have some lulus, because heat and humidity give air just the kind of instability a thunderstorm feeds on. But perhaps the tornado belt of the American Midwest gives rise to the worst, when that warm, moist air from the Mexican

Gulf starts flirting with a cold, dry air mass down from the Arctic. Really huge hail—I mean orange- or even grapefruit-size—and tornadoes (both the products of really severe thunderstorms) are almost exclusively a phenomenon of the Midwest. Yet the airline pilot who is the most experienced of all my flying acquaintances says the most turbulent thunderstorm he ever met was near Montélimar, in France's Rhône Valley, while flying a Lockheed 1049 whose radar was out of commission. So, having encountered the thing, what does one do? I asked.

"Tighten your harness, slow down to rough-air speed, lower your seat, and turn up the cockpit lighting to reduce the chances of being temporarily dazzled by lightning. Fly a constant attitude, paying not too much attention to airspeed and altitude excursions as the vertical gusts get hold of you." He added: "Of course, you plan to stay out of thunder-bumpers. But if you fly instruments regularly, sooner or later you'll blunder into one."

After you've been flying for some years, you develop a familiarity with weather that your nearest and dearest come to take for granted. "What's it going to do today?" they ask you with casual confidence in the morning, and though you are sometimes wrong, you are usually close enough. If you are a private pilot, your forecasting ability tends to improve as the weekend approaches and you give the weather more attention. Your Saturday morning forecasts are infinitely sharper than, say, your workaday Tuesday prognostications. All it takes is a glance at the morning paper or the TV weather map, and a look at the sky to see whether the cold front has gone by or not. Although wind may blow from any point of the compass, in temperate climes weather moves steadily in one direction, generally west to east over North America and Europe. You quickly learn that weather generally travels about two hundred miles a day in summer and three hundred in winter, so you look at the weather map that far upwind of where you live, and you know that what they're getting now you can expect in so many hours' time. The low-pressure systems, at least, move quite steadily; highs tend to hover, to edge slowly north or south, and are just as likely to dissipate *in situ* as finally to move off elsewhere. So if

there's a high sitting to the south, and another forming way north, with a small low-pressure area moving between them, and if the isobars (lines of equal atmospheric pressure) on the map are getting close together as the low is squeezed between the two highs, you say, "Going to be windy," and it usually is. Very nourishing to the ego.

High pressure, you learn, produces different weather in winter and summer. In summer, days are longer than nights, so the clear skies that a high brings allow the sun to warm up the land, and though the atmosphere may be hazy after dawn, you know it will be hot, with light winds, by noon. In winter, clear skies at night let the earth radiate heat off into space. Since nights then are longer than days, and the sun is a less effective heater because it is low in the sky, winter highs bring colder temperatures and often fog. England's winter fogs achieve maximum density when a cold high moves in soon after a mild low has deposited a lot of rain and left the landscape sodden. These special conditions are most likely to develop around the first two weeks in December.

Weather is generally better in every way in summer than in winter. This isn't just an illusion brought on by warmer temperatures. In summer the cloud ceiling is above 1,000 feet and the visibility better than three miles (the minimums for visual flight) at least ninety percent of the time over almost the entire United States. The few exceptions are coastal southern California (Los Angeles smog), San Francisco (sea fog), the region around Seattle, and a few areas along the eastern seaboard, such as the Carolinas and Massachusetts. But in winter much of the eastern half of the country, the northwest, and even California has "instrument" weather (below 1,000 feet and three miles) for at least a quarter of the time. Mind you, these are *averages*. There are winter days in New England, soon after a cold front has gone by, when there is no cloud and you can see for a hundred miles; fine flying, except that the strong northeast wind will make for turbulence low down and over the hills.

Fascinating to a pilot is the way weather—even the same type of weather—changes around the world. For instance, a North American pilot may

"Any student pilot can tell you that relating what you see on the earth below with what is printed on a map is quite a trick; it comes with practice." Left: Pilot checks landscape. Below I: Rockwell Commander 112A crosses well-defined highway. Below: S-curves in river should help locate airplane over Pennsylvania.

have read in textbooks that warm fronts can bring ceilings and visibilities of zero, but never have experienced one doing anything like this. But he will if he flies in western Europe. The pilot who flies in temperate latitudes is concerned with the low clouds and poor visibility associated with lows, and with fronts, warm or cold. If he flies at night he may occasionally run the risk of radiation fog forming at his destination while he is en route. If his airports of departure and destination do not offer a choice of runways on windy days he may be concerned about the wind's strength and direction, lest its velocity be above his aircraft's cross-wind limits for takeoff and landing. In summer he may have cause to ponder the chances of thunderstorms along his route. They can build up from almost nothing in the course of a couple of hours.

The piston pilot who is going to fly by instrument rules, maybe in cloud, and perhaps at altitudes determined by air-traffic requirements rather than by his own choice, will also want to know the likelihood of icing. What is the freezing level? What are the chances of clouds above that level being ice producers? And is there any risk of thunderstorms not clearly visible in advance, but embedded in an ordinary cloud mass? (Icing hardly bothers the jet pilot, who can climb up and down through icing levels in a trice.)

The hang-gliding enthusiast has simpler weather needs: just that the wind be blowing across his favorite slope at about the right speed.

The Arctic pilot—there's a lot of flying in the Arctic these days—has weather considerations all his own, and never mind the task of getting his engine started when it's so-many below. His problems are mainly restrictions on visibility: snow or blowing snow during landing; special kinds of fog (advection fogs along the coast, ice fogs on those exceedingly cold calm winter days, and steam fog over the ocean); and "whiteout," a strange optical effect that occurs when snow-covered landscape and low cloud of uniform thickness diffuse the light from a low sun. Earth and sky seem to merge into one great blank bowl. Depth perception is lost. The horizon disappears and dark objects on the ground seem to float in space. In a whiteout you can fly into the snow-covered ground without ever seeing it. A similar condition, "glassy water," plagues the seaplane pilot. From aloft, when there's a thick haze and no wind, you just cannot see the water's surface.

The flier in the tropics may have to calculate the loss of engine power due to high air temperature, particularly if the airport is high as well. Between them, hot and high can double your takeoff run. Tropical weather is less variable, more predictable than elsewhere. It is best in the forenoon, before the heavy showers produced by cumulus build-ups. Visibility usually is good, despite the humidity, but less good in the rainy season. The atmosphere is likely to be more often calm than windy, except in the hurricane season. (Or typhoon, or cyclone; they're all the same phenomenon.) These blows are so furious that nowadays they are tracked and reported with enormous care from weather satellites. In circumnavigating one of these the word in the Northern Hemisphere is to leave the storm on your left, when you can expect following winds. How violent are they? By definition, anything more than 64 knots is "hurricane force." A top of 175 knots has been measured. Usually the anemometers break long before then.

Tropical deserts like the Sahara sometimes give poor visibility in sandstorms, but otherwise flying weather is generally fair—except for the wretched heat turbulence once the sun is well up.

The jet airliner captain embarking on a transatlantic crossing has quite different aspects of weather to contemplate. Fronts, lows, or surface weather are of little consequence. At 35,000 or 45,000 feet he's "on top" and his skies are mostly clear. He will check the forecast for his destination, but with less concern than a small-airplane pilot. Seven or eight hours will elapse before he arrives there, and conditions will very likely change by then anyway. His big jet can handle most weather, and most cross winds for landing. His destination is likely to have a full Instrument Landing System, with which he can accommodate any weather down to 200 feet cloudbase and half a mile visibility. Or, if he and the ILS are cleared to "Cat. Two," 100 feet and a quarter-mile. (British Airways Tridents can make fully automatic landings in even dense fog.)

On the Numbers: Tilted plane approaching level field will achieve a smooth and professional landing if it touches down along centerline near numbers indicating compass direction runway is aligned with.

No, his first concern is "winds aloft," for he flies at jet-stream altitudes, among the rushing, invisible rivers of high-speed air that twist around the earth. Their velocities can be 50, 100, 150 knots, or even more. Even a modest fifty-knot wind "on the nose" (head-on) can give the pilot an extra 350 nautical miles—say, an extra fifty minutes—to fly on a seven-hour crossing. If he is lucky enough to ride a jet-stream tailwind from North America to Europe, flying time will be four or five hours instead of the usual seven.

He will also pay attention to the turbulence part of his forecast. He may see hardly a cloud at 35,000 feet, but his passengers can still experience a rough ride from clear-air turbulence, particularly in the narrow region of wind shear that surrounds the core of a jet stream.

If he should face the prospect of diversion to an airport other than his intended one, it is delay rather than distance that matters to the jet captain. If fog at London forces him to divert to Manchester, or Paris, Amsterdam, or Frankfurt, the delay—at 600 mph—will be minimal. Going the other way, however, a delay for traffic headed into JFK because of congestion or weather may become a matter for concern. He sits there, going round and round in a holding pattern and studying his rate of fuel consumption ever more intently, lest the moment come when he must forget New York and proceed to Boston or Philadelphia, somewhere they can let him land as

69

Weather: Haze, photographed
over Missouri at 9,500 ft (2,896 m), is
a concentration of salt or other
dry particles occurring in stable air.
Downward visibility is impaired,
landing into sun through haze is tricky.
It will not dissipate, must
be dispersed by air movement.

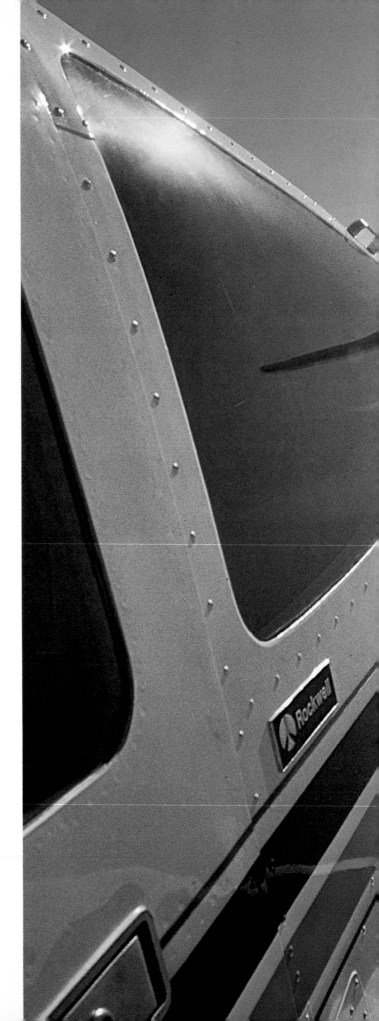

soon as he arrives. Jets are efficient users of fuel when cruising at speed at altitude. The miles seem to melt away far faster than the pounds of kerosene remaining. But throttled back, jet engines are inefficient, and at low altitudes their thirst is dire. American law, for example, requires you to carry enough fuel to proceed to your destination, then proceed to your planned alternate destination and still be able to fly another forty-five minutes. But few pilots would be happy to find they were operating at that bare minimum.

The Concorde captain is less concerned with winds at his cruising altitudes. At 55,000 or 65,000 feet he is above the jet streams, and even a pretty strong wind does not bother him. He is doing better than 1,000 knots, and in any event is only up there for three hours or so. What concerns him is the *temperature* of the air at his altitude, for a small temperature variation makes a big difference in his fuel consumption. In addition, he is cruising at Mach 2.02, and his overspeed warning sounds at Mach 2.04—*close*. You will remember that Mach is a function of air temperature as well as speed. It takes only a 4-degree C change in temperature to get him to 2.04 and a warning to slow down. And temperature shears—sudden changes—of 12 degrees C are not unusual at high altitudes in the tropics.

Pilots get weather information in a variety of ways. Ordinary newspaper or television forecasts offer a broad picture. In most parts of the world there are telephone numbers you can call to get a recorded announcement on the latest aviation weather. You can telephone the Weather Bureau ("Met Office" in Europe) and speak to a forecaster about your special needs. Larger airports have self-briefing rooms for pilots, with teleprinter readouts and facsimile weather maps on display. Once airborne, you can listen to the radio station you are using for navigation. At certain times each hour the Morse or voice ident of the station is replaced by weather information. Or you can call a flight-service station for information. Certain radio beacons transmit continuously a tape recording of weather details. Notable among these are certain beacons across the United States which broadcast a synopsis of the weather in their area, plus the latest "ac-

70

Weather: Big, cottony cumulus clouds occur at low
levels, but may develop vertically into huge towers, as
these below are threatening to do. That means
storms and turbulence for fliers—as do mamma clouds (r).
Inexperienced pilots seeking to avoid entering these
sometimes try to go underneath, where downdrafts
and hail can be even more dangerous.

tuals" from quite a long list of airports. Because this information is endlessly repeated, you have a chance to pick up any details you missed as the tape comes round again. (I use a little portable transistor radio to listen to one of these transcribed weather-broadcasting beacons from my city apartment before I leave for the airport. It often provides all the weather information I need.)

With this wealth of information so freely available, it is pathetic that so many private pilots who get in trouble flying into weather they can't handle turn out not to have checked the weather at all before they took off. Why? Too lazy, or impatient, or foolishly proud, one supposes. Most often the forecast they couldn't be bothered to check would have warned them quite accurately what awaited; and usually their airplanes have the necessary gear for instrument flight in poor weather, even though they have not acquired an instrument-flying capability.

How do we find our way about in the air? Me, in my old biplane, I have no electrics or radio aids of any kind, so I use the three basic functions of all air navigation: a map, from which I pick out and identify details such as highways, rivers, railroads, coastlines, cities, mountains; a compass, for flying a straight line; and a watch, with which to estimate how far I have gone. (This is dead reckoning.) Even this time function I do not use often, but it comes in handy when crossing the sea, so I can estimate when I will sight the opposite coast, and also on long journeys, so I can be sure to reach my planned destination before exhausting my three-hour fuel allowance. Navigation this simple got Lindbergh from New York to Paris in 1927, and can still get you pretty much wherever you want to go in fair weather, except that without radio you are obliged to avoid many pockets of "controlled air space" around the principal airports and military air bases. No matter; these are marked on your map.

Any student pilot can tell you that relating what you see on the earth below with what is printed on a map is quite a trick; it comes with practice. Eventually you will draw your planned track across a half- or quarter-million scale map with exceptional accuracy. With concentration you can make the whole flight without wandering more than

73

Weather: Rising bulge of cumulus congestus
penetrates stratocumulus, with high-level cirrus
(composed of ice crystals) in background.
Flying will be rough in cumulus, smoother above,
though below cirrus. Right: Cumulonimbus
is bad news, promises thunderstorm with
lightning, turbulence, and hail as extras.

quency Omni-Range, abbreviated to VOR or Omni. the ground beacons transmit a double signal with a phase difference that varies according to the direction in which the signal is radiated through the 360 degrees around the ground station. Your airplane has a receiver that measures this phase difference and displays the result to you as a radial, or bearing, to or from the station. These VORs are generally used to define airways, and most flying today seems to consist of hopping from one VOR to the next. VOR is precise, but of limited range. Of course, you can tune one off to one side of your track to give a cross-radial and thereby a position fix, but a better aid is Distance Measuring Equipment (DME), usually colocated with a VOR on the ground. The

DME set in your airplane transmits to the ground station, triggering it to reply. DME measures the tiny fraction of a second it takes for the reply to be received, and displays this to you as your distance to that ground station. Radial (bearing) and distance are all you need to know to tell you exactly where you are at any time. And VOR/DME is the classic navigation system used by the airlines and everyone else in every technologically advanced nation. (NDBs and ADF are still important in the Third World.) The only snag is that it obliges everyone to tread the same beaten tracks across the sky, from one VORTAC (the name for colocated VOR/DME ground stations) to the next, with some consequent congestion of traffic. So the avionics-equipment man-

ufacturers came up with Area Navigation (RNav), a system that combines a simple electronic computer with the basic VOR and DME receivers, and effectively enables you to "create" phantom ground stations (called "way points") wherever you want, and to lay down an airway for yourself wherever you like, as long as there are VORTACs close enough for your RNav gear to define it. RNav is of course much more complex and expensive than the old simple VOR/DME, and the avionics manufacturers are crying all the way to the bank. Pilots grumble that, like all avionics, the damn stuff isn't reliable. It doesn't work half the time and you can never find anyone to fix it properly, and there is a good deal of truth in their plaint. So, for safety's sake, airplanes (and air-

liners in particular) tend to carry at least two of everything.

Or even three. Installation of three Inertial Navigation Systems is not unusual, and they are extremely expensive—at least $100,000 each. The INS is an offshoot of the space program, having been originally developed to navigate ICBMs to their targets. The INS uses accelerometers and spinning gyros to measure every change of motion of a craft, however minute these may be. So before you start taxiing, you dial into the INS the coordinates of where you are at the airport terminal, and thereafter it tells you wherever you are, in latitude and longitude. It also tells you your groundspeed, and it does this so accurately that Concorde and 747 pilots use it to

79

check their speed while taxiing for takeoff; their cockpit is high enough off the ground that they find it hard to estimate their modest speed down the taxiways just by looking out. INS is splendidly independent of ground stations, and therefore a superb aid to over-ocean flying, when you are out of range of the VOR/DME network. (There are other such navigation aids: Loran, Doppler, Decca, Astro, Omega VLF, but I'll refer you to the textbooks for details on them.) One other piece of electronic equipment I will touch on: the transponder. This distinguishes your blip on the ground controller's radar screen from any other traffic he may be "painting." You set your transponder to a code he tells you, and it may also tell him what your altitude is.

All these nav aids are basically for en-route navigation. Though they can be used for approaching an airport, none of them is really precise enough to be ideal for this. So all the world's major airports have Instrument Landing Systems which radiate ultra-high-frequency signals in a cone-shaped pattern up into the sky at a slight angle from the start of the runway. One by one, aircraft descend through the center of this cone to land. The horizontal signal is the Localizer, which tells you with great accuracy whether you are left or right of the ideal path; the vertical signal is the Glide Slope, which tells you whether you are descending at the right angle. A third signal, called the Outer Marker, is radiated in a fan-shaped pattern across the cone; it marks a point at which you should be firmly established on your ILS "approach." (There may also be Middle and Inner Markers closer to the runway's threshold.) There is usually an old-fashioned low-power NDB, typically colocated with the Outer Marker to help you to aim for it initially. This NDB is called the Locator.

An essential part of every ILS is a set of really bright runway lights to aid your transition from instrument indications inside your cockpit to outside visual cues for landing. The classic ILS presentation to pilots is two crossed needles: keep them crossed in the center and you are on course, on glide slope. Pilots of air transport and executive jet aircraft have more sophisticated instrument displays, called Flight Directors. Here a computer is employed to di-

gest the raw radio information and present it to the pilot as smoothed-out steering commands, easier to follow.

Come fly an ILS with me. Having come off the main higher-altitude airway system, we have been cleared by the (radar) approach controller to descend in stages to our initial approach altitude, typically some 1,800 feet above the ground over flat terrain. He will usually give us radar vectors (headings to fly) to intercept the ILS, but there is also the Locator NDB to help us, if needed. As we come up at a slight angle on the centerline of the ILS, our Localizer needle centers itself, and we turn onto our final approach heading—making a slight allowance for any cross wind that may be blowing. At first our Glide Slope needle shows that we are *under* the glide slope; indeed, at this safe altitude we plan to intercept it from underneath. As we come up on the Outer Marker our Glide Slope needle descends till it reaches the center of the dial just as the *beep-beep* code of the Fan Marker itself sounds in our earphones. As we lower the landing gear and start to descend, usually at around 500 feet a minute, we cross-check our instrument indications. The Glide Slope needle is centered as we cross the marker, our altitude just leaving 1,800 feet and the ADF needle just starting to point back as the Outer Marker passes below and behind us. From here on, flying by reference to the ILS indications as well as our main flight instruments, we make small corrections with the aircraft's controls to keep the two needles centered. We start to glance ahead into the murk beyond the windshield. Before we have descended to our Minimum Decision Altitude or Decision Height we must have those runway lights clearly in view. This critical height is low, typically only 200 feet, and if we are still in the clouds even then we must make a missed approach—pour on the coal and climb away for either another approach or else a diversion to an airfield (most likely our chosen "alternate") where the weather is known to be better.

But ordinarily the lights are there, clear enough, in plenty of time, and we continue our descent by their aid to the runway threshold, throttle back, and land. The accuracy with which an ILS can deliver you to the runway threshold is uncanny.

80

Most aircraft, by the way, can have their autopilots coupled to the ILS receiver to fly an automatic approach. The human pilot does have to take over for the actual landing, though an increasing number of airliners can make fully automatic landings. Here the pilot's job is one of monitoring, watching the automatic systems operation to ensure that they are functioning correctly. So far the avionics manufacturers have only been able to achieve the sort of reliability required of their electronic machines by "redundancy." The British Autoland systems, for example, have three of everything, with any two gadgets strong enough to overpower the third should it develop wild notions of its own. (The British also fly with three human pilots.) Large jet aircraft are automated to a degree that is astonishing, almost disturbing, to an outsider. In a little Cessna you are actually turning a butterfly valve in the carburetor when you move the throttle, and actually moving the controls by wires when you move the yoke, but you have nothing like such direct control in a jet. You are merely opening hydraulic valves and allowing fluid to flow down pipes and move pistons attached to the controls. When you move the power lever you are merely conveying to a complex assembly of fuel metering and pumping equipment what you would like it to do. In its own good time it will do it, but at its pace, not yours.

The captain of a 747 may put his bird on autopilot on the climb-out from London and leave the automatics doing the donkey work till touch-down at Los Angeles. He is still very much in charge, but exercising his control by dialing instructions to the autopilot rather than by hand-flying the aircraft. But instrument flight *training* always proceeds as though autopilots had never been invented. With good reason, for they can always fail.

Getting an instrument rating is the most difficult hurdle in anyone's flying career—so difficult, in fact, that most private pilots never even attempt it. First there is a blizzard of book work to be mastered —minutiae of aviation law and flying procedures to be studied in such close detail as to approach absurdity. You can do it through formal ground-school classes, correspondence courses, home study, or tape-and-slide aid devices. I used all these methods, and it still took me six months of evenings, weekends, and occasional half-days off from work. The FAA exam is of the multiple-choice sort. Typically, two of the answers are fairly obvious "dogs"; one is

81

Mountains and ice confront lightplane
pilots with special problems. Thinner air at high altitudes
means less lift and less performance from plane.
Ice (or snow) may combine with low cloud to
create dangerous "whiteout" which erases horizon,
destroys depth perception. Cloud-shadowed water (r)
still has definition for seaplane pilot, but "glassy water,"
whose surface cannot be seen or judged, is a danger.

"Flying is . . . for those who relish the difficulties,
the book work of learning, the cost of flying time, the hard
slog of that practice that in the end maketh perfect."
Student at Learjet school studies hydraulic system (top),
cockpit simulator, and operations manual.

a deceiver; one is correct. You are given eight hours to complete it and you need seventy percent to pass.

Then follow some fifteen or twenty hours of learning procedures in a simulator, with an inky mechanical crab tracing your erratic path across a tracking table outside your cabin. Then the flying with an instructor. You wear a hood that allows you to see the instruments but nothing outside the airplane. You will suffer modest despair at the complexity and seeming impossibility of it all. One instrument-rating student, Gordon Baxter, a local radio disc jockey and a writer for *Flying* magazine, says that his listeners used to receive their news and music mixed up with "witty recitations from the Federal Air Regulations" when he was studying. And once, when under the hood and seemingly faring quite well, he so exasperated his instructor by making a hash of a practice ILS approach that the teacher shouted: "Now what the hell did you do *that* for?" "Because I'm a damned student, that's why!" said Bax, and anyone who has ever been a damned instrument student will know how he felt. It is usually at about this point in your training that the instructor decides to show you how to do it. Under his deft hands your own rash darts and sallies on the airplane's controls are replaced by smooth, almost imperceptibly tiny motions, and he'll fly an ILS for you with both needles seemingly *nailed* together in the center of the instrument. Silently you curse him, and then yourself. But you end by thinking, "If he can do it, in the end I'll learn." If you do despair and abandon training, perhaps owing to the pressures of other work, you may console yourself with the thought that you can always come back and finish later. Plenty do.

If you plan to be a professional pilot, you *must* earn an instrument rating, unless you are content to be a flying instructor or duster pilot all your days. Transport flying is where the big money is for a commercial pilot. And if you become an instrument-rated *private* pilot, it is a fine feeling to know that the eighty percent of pilots who aren't will never know more than twenty percent of what you do about how to fly. "I realized," wrote Gordon Baxter, "that flying may be a great game, but it can only be survived by everybody knowing and playing by the same rules. Yet eight out of ten of the players are only required to know about one-fourth of the rules. Freedom of the skies be damned; this is madness.

"I may become insufferable about this, like a reformed drunk busting into saloons and slapping the drinks out of everybody's hands, but for the first time in my life I don't feel too guilty about calling myself a pilot in front of other pilots. With the ink still wet on my ticket, I'm no instrument pilot yet. . . . But I'm working at it. And I have never worked so hard for, or been so proud of, anything in my life."

Flying is many things to many people, and most fliers have nothing in common save their acquired skills. To the 747 captain who began as a co-pilot on DC-3s it is a very pleasant way of earning $80,000 a year, and one that may have taken him to every strange land on earth. I remember arranging a meeting with one such old-timer in a hotel lobby. How would I know him, I inquired. "Oh, you'll have no difficulty recognizing me," he said. "I'm just kind of an old bastard." And a most courteous, entertaining, and worldly-wise one, too. To the eighteen-year-old with a new private certificate it is something that, though he—or she—may not know it, has already raised him above his contemporaries whose sights are set no higher than a used Corvette. To the air-force fighter pilot it is comradeship, enormous responsibility gained young, better than working in a lawyer's office, and all he ever wanted to be. To the glider pilot, a breed apart from power fliers, it is simply reincarnation as a bird. To the ordinary private flier it is weekends in a world away from workaday cares, marvelous relaxation—the best—and something that gives Friday night's television weather forecast real significance.

The most childlike part of my pleasure in flying comes from using the radio. I love the simple coded speech, conversation reduced to its absolute minimum. (Why does Hollywood always make such hash of radio conversation? Did anyone besides a scriptwriter ever say, "Come in, please," *ever*?) I love the poetry of the international phonetic alphabet: Alpha, Bravo, Charlie, Delta. Did you know that this form is its second transmutation? It was Able,

85

Night Landing: Bright pattern on runway
lights gives pilot visual cues for where and
when to touch down. Glowing dials in
cockpit indicate plane is on course for ILS landing.
At bottom, below, plane is on ground.

La Guardia Tower, New York: From
here controllers organize the flying on and
off one of the world's busiest airfields
for everything from lightplanes to jet airliners.

Baker, Charlie, Dog in World War II, and Ack, Beer, Charlie, Don in the trenches of Flanders. (Charlie is seemingly immortal, even if his comrades aren't.) The radio enables you to call up spirits out of the ether, to serve you as you will. "Hometown Municipal," you transmit, as soon as your engine is running, "November Two One Five Three X-ray, VFR to Lakeville Junction. Taxi, please. Over." And back comes this genie whose face you may never see, bidding you good morning, and catering most cheerfully to your most modest whim.

Do you wish an intersection takeoff, rather than trundling to the runway's far end? He will allow it. Do you want to make a straight-out departure, rather than making the ritual turns, 90 degrees left, then 45 right? Most likely he will approve it. Such power you have, simply by depressing your microphone switch! And you also try to make his life simple: on your return to his field you listen on his frequency for a minute before calling, thus learning from his instructions to others which runway he is using, what the wind direction and speed are, what pressure you should set the altimeter to. "Hometown, Two One Five Three X, five west with the numbers," you say, to save him yet another recitation of his litany. "FivethreeXcalldownwind," he says in one breath, and so you do, adding that you are behind a Cherokee. "Rock your wings," he'll call if he has any doubt which one you are. La Guardia, Van Nuys, Washington National, O'Hare, these are the heavily congested fields where the controllers' chatter issues faster than a carnival barker's spiel. Is it true that after a short stint in such a bedlam FAA controllers are put out to pasture at some quiet country field chosen for its restful somnolence and lack of traffic? It would seem no more than right.

Sometimes controllers seem motivated by a private mischief all their own. "Report Salmon Lake," one will tell you, though you are an obvious stranger to his air space, as if he expected you to be able to tell from 1,500 feet which of the thirteen small expanses of water below you still had any fish in it. "Report the stacks"—and there are three power stations to choose from. "Report the drive-in" —but tonight's double bill hasn't yet begun.

Regional accents can also have you scratching

88

your head. You hear some astonishing Brooklynese on the airways around New York. "See fool, solsabod, an confuh mun arefly intray nin aria," a new instructor claimed he was once asked on taxiing out for his first flight with a student down in the Caribbean. ("Say your fuel endurance, souls on board, and confirm that it's a one-hour flight in the training area.")

Then there are the anonymous interrupters. One gusty afternoon I suggested to the tower at New York's Westchester County Airport that he consider changing the runway, since the cross wind was something like 90 degrees. "Learn how to fly," advised some other wise pilot before the tower could answer. ("Sir, I'm trying.") And in the days when I was an RAF controller, we had a jet-fighter aircraft whose gear would not lower. But he had a whole lot of fuel, so we foamed the runway and got out the fire crew and the ambulance, and advised him at length on how to proceed on 121.5 megaHertz, the international distress frequency. "Okay. I'm on finals. Will try to touch down as slow as possible," he said, and did so, slithering at last to a moist stop, unharmed. Silence followed on 121.5, broken eventually by an anguished cry from a USAF eavesdropper: "RAF, for gawd's sake, tell us: *what happened to that guy?*"

I myself am a somewhat less-than-perfect pilot. I like to think it's because my work prevents me from getting enough practice to stay truly current, a predicament that snares a great number of private pilots. I try to disguise my rustiness by "cheating." This is a technique I learned from the instructor with whom I worked for an instrument rating. One day, noting that I was a hundred feet lower than I should have been, I quickly pulled up to my proper altitude and apologized for my lapse. "Why did you do that?" he asked. "You should have said nothing, and snuck back up very gently. I never would have noticed. You've got to cheat a little to be smooth." Now I cheat a little whenever possible. You know how the first thing you do when you start taxiing is to apply the brakes to be sure they're working? I used to do this so you would really know what I was doing. Now, as gently as possible, I make that first power reduction after takeoff slowly

and smoothly. I raise and lower flaps slowly, ready on the yoke for the accompanying trim change. Best cheat of all is that of a British professional pilot friend who, when he fumbles or goofs over something on an instrument-rating check ride, asks his examiner to look out the window and check for ice on the wings. (You are allowed to ask your examiner to perform some copilot duties.) They may be in a perfectly clear sky with the nearest icing weather a thousand miles away, but the poor examiner has to go through the motions of checking for ice. By the time he's tumbled to the trick, my friend has found the approach gate he was looking for, or reset the ILS to the right frequency, or whatever. And he always passes the check.

Flying does demand a light touch. It is, after all, the only means of transportation that employs an invisible medium, and that requires a transition from and to a quite different medium to be made at high speed at the beginning and end of the trip. So hours of one's early training are spent strengthening coordination between hand and eye, hand and hand, hand and foot. Lazy eights and pylon eights teach you to slowly f-e-e-d in bank while e-a-s-i-n-g off elevator, and so on. Aerobatic pilots are probably the only people who ever get to use the full range of control deflections possible, and even they are far from violent with it. Most piloting is a matter of small, delicate control inputs, allied with a good deal of thinking about what to do next. It is an excellent pastime for those with a touch of the perfectionist in their make-up.

Such a tactile skill is perhaps most easily acquired in one's teens. In most countries you may legally fly solo in a powered plane from your seventeenth birthday, and in many countries earlier still in gliders. But many people do not take up flying till middle age, when the kids are grown up and they at last have some money to spend on themselves. Mr. Theo Ceasar of Auckland, New Zealand, took it up at seventy-three and, with the encouragement of his wife, earned his private license at seventy-four. Mrs. Ceasar herself went solo for the first time at sixty-eight.

So you see, it isn't really aerodynamics that keeps a plane up. It's enthusiasm.

Aerobatics

No matter how long and hard you struggle, you can never really get your harness tight enough. You can pull in all the straps until you can hardly breathe, but once aloft, as soon as you roll over upside-down, you seem invariably to rise off the seat and hang, as they say, *loose*. Uncomfortable. I've seen some Pittses with a new-type harness that you tighten against a ratchet, and the pilots do claim you can get really tight with these; but my thirty-five-year-old Bücker Jungmeister has an old-fashioned linen-and-buckle strap arrangement, and I can never get it tight enough. So what I do is tighten it as far as I can before even starting to taxi, and then have another go just before takeoff. It's a standard five-point aerobatic harness, one over each shoulder, two around the waist, and the fifth coming up between your legs. This holds you to the seat. I've also installed an extra backup lap strap that holds me *and* the seat to the airplane, just in case anything should let go or come undone—what in Yorkshire they call the "belt *and* braces" approach.

The other pre-takeoff checks are conventional enough, but I pay particular attention to ensuring that the controls are free and giving their full range of movement; you'll see why in a moment. I even disconnect the rudder from the tail-wheel steering to be sure it waggles as far as it should. Okay: line up, full power, raise the tail, two bounces and we're airborne (160 hp in a light airframe) and climbing steeply at a bare minimum 80 kilometers per hour to an altitude of about 200 feet. Here I level off and accelerate slowly to 170 km/h. Once that magic figure is reached, I start raising the nose—stick coming back, all the way back—kick on full right rudder and hold both control deflections as we spin through upside-down. About 45 degrees before we are back to level I relax the back pressure and give a sharp dab of opposite (left) rudder to stop the snap roll. Recovery takes, quite literally, a single second. We finish in level flight, same height we started, but with the airspeed now suddenly back to around 100 km/h. And we carry on, climbing away, as though nothing at all had happened.

The Bücker is the only airplane I'd care to do this in, the only aerobatic type with such a quick, deft, utterly reliable snap roll that you know it'll go

(and stop) just as predictably every time. You wouldn't, for instance, want the snap to stop or run out of steam just as you hit inverted, for at that moment you are pointing steeply down—not a good thing at a mere 200 feet. You *can* do it even lower than 200 feet, but it tends to get the man in the control tower going and demanding a word with you afterward. ("Just what the hell did you think you were doing?" Nothing you can say then but "Very sorry, Mister.") And in any case, I am a devout coward and a very amateur Sunday aerobatic pilot who doesn't get the practice he should, and 200 feet is quite low enough for a snap roll to feel (and look) exciting. Mind you, there was a man named "Buzz" Cantacuzene, a Rumanian prince, who had a Jungmeister, and he would snap it literally right down to the deck. He had a trick, almost the signature to his aerobatic performances, which he began by seeming to have finished his display and to be making a landing approach. If you watched closely, though, you could see that he was going a little fast to land, even though he was throttled back. Then he'd touch his wheels momentarily on the turf, apply full power, pull the nose up to about 45 degrees above the horizon, and snap right there, his rotating wing-tips sometimes as little as 20 feet above the ground. Then he'd close the throttle and finish the landing. I once saw him do this astonishing trick three times in a row, along the front of the crowd at the national air races meeting at Coventry, England. Cantacuzene was a natural showman and something of a clown. He'd wander over to the crowd line before he was due to fly at an air show, select the prettiest girl there, and say, "You are beautiful woman, I kees you," and promptly *kees* her. And when doing his mag checks during an engine run he would momentarily switch off both magnetos and in the instant silence cover his face with his hands in mock despair and glower at his mechanic, an old-fashioned Englishman who called him "Old Matey" and pretended to be totally unimpressed by Buzz. Buzz Cantacuzene was married to a ballet dancer, I think, and against all prognostications he died peacefully in his bed. He was the greatest Jungmeister exponent in Europe. Bevo Howard was the great Bücker pilot in the United States. He died less happily—dur-

Preceding pages: Smoke plume defines
route of Pitts Special performing double
snap roll at an air show. Below:
Portrait of author near top of vertical
climb in execution of hammerhead.

ing a show, he ran out of gas while upside-down and
hit a tree.

Which brings me to a point I have been won-
dering quite how to get round to: that aerobatics is
a dangerous sport. I took it up in 1960 and a sad
number of the aerobatic pilots I competed against in
contests then, or have otherwise met since, are now
in their graves. I can think of Neville Browning and
Charles Boddington in England; Bevo, Bill Adams,
Harold Krier, Bob Schnuerle, and others in the
United States; Gerard Verette of France; Vladimir
Martemianov of the Soviet Union. And of those who
are, happily, still with us, several have had some
close shaves. None closer, perhaps, than Neil Wil-
liams, a quiet Welshman who has been the British

national champion eleven times (in itself a unique
record), but who has demolished two airplanes
along the way. In one, a Stampe biplane, he spun in
during an air show. ("He's taking it behind the hangar!" cried the commentator as Neil disappeared;
and as a big cloud of dust blew up from where he
went in, just out of sight: ". . . and he's left it behind
the hangar!") On another occasion Neil had a
midair structural failure in a Zlin monoplane: one
wing started to fold upward at the root. So Neil,
surely acting more calmly than he felt (he wore no
parachute), rolled onto his back and ran an inverted
approach to land, rolling upright only at the last moment before landing, as the wing finally gave way.

I am not saying that the sport of aerobatics is

93

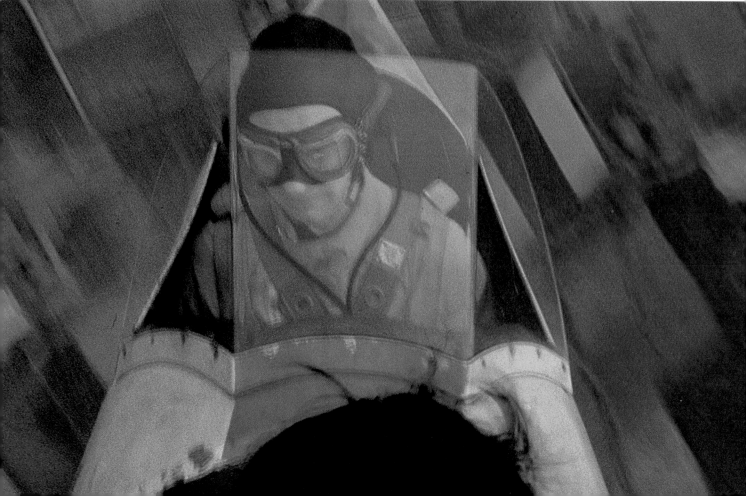

Bücker Jungmeister (l) and Stampe SV-4 (below). Sunburst paint schemes are traditional for aerobatic planes —and so are engine modifications to assure constant oil and fuel in inverted flight. Below l: Author's front-cockpit photograph of Stampe on its way through a snap roll. Stampe is a Belgian design.

any more dangerous than, say, motor racing, or mountaineering, or parachuting, or skin diving, but it is definitely in that class of inherently hazardous pastimes.

So we'll both wear parachutes and we'll climb to a goodly altitude, five or six thousand, before I show you the basic maneuvers. Straps tight? Really tight? First, a couple of "clearing turns," 90 degrees left, 90 degrees right, to scan for other traffic in our area. Clear? Then lower the nose and let the speed build up for a simple loop. Back with the stick, throttle fully open as we pull up through the vertical, start relaxing the back pressure as we go upside-down over the top (when we are starting to sit lightly on our seats but still, through centrifugal force, not actually hanging in our harnesses), and gently feel our way down the back side of the loop, returning to level flight at much the same airspeed we had when we left it. Afterward the accelerometer will probably show a maximum of three g reached— enough to make you feel a trifle leaden and pushed down into your seat, but nothing too uncomfortable. Except for the g, the sensation is much as though the earth and sky had taken a tumble around us. You don't need to be an Einstein to appreciate that motion is relative. This g, by the way, is a measure of acceleration, or the force of gravity. One g is by definition the force of the earth's gravity we all live under. If your normal weight at one g is 170 pounds (76.5 kg), you will weigh 340 pounds (153 kg) at two g, 510 pounds (229.5 kg) at three g. So you can see at once why aerobatic airplanes must be espe-

cially strongly constructed.

What next? Some rolls, I think. The easiest kind is the barrel roll: simply pull the nose up well above the horizon and push the stick all the way over to one side. You should finish once more in level flight, having had positive g all the way around—that is, sufficient centrifugal force to keep you touching your seat even when upside-down. The maneuver gets its name from the corkscrewlike pattern—fancifully supposed to be barrel-shaped— the airplane describes in the sky while doing it. My Bücker does beautiful barrel rolls. Sometimes I enliven a dull cross-country flight by doing a whole series of them, one left, one right, one after the other.

The true slow roll is not nearly so easy, and demands some practice from the novice. First, make sure your harness is *really* tight, for in a moment you will be hanging in it, and while a loose harness is not dangerous, it is disconcerting (particularly in an open cockpit) to find yourself upside-down a mile high with only an arrangement of loose straps between you and a long drop.

Start in level flight, or a hair nose-high, with full throttle and an airspeed maybe 10 knots above normal cruise. Apply full aileron deflection one way or the other; as you roll through a vertical bank, add top rudder—quite a lot of top rudder—to prevent the nose (which contains the heavy engine, remember) from falling away. As you reach inverted flight, you will suddenly be hanging upside-down— perhaps for the first time. At this point you very

95

likely will be strongly tempted not to complete the roll and to try to restore normal positive g—and that comforting pressure of the seat against your bottom—by pulling back the stick and diving through a half-loop to comfortable level flight. Don't, no matter what, give in to the impulse. Because you will consume an enormous amount of sky in your recovery—perhaps more than is available—and you are likely to achieve an airspeed well in excess of "redline," that speed determined by your airplane's manufacturer to be the safe maximum. Did you realize that most light aircraft are test-flown by their builders to airspeeds only ten or maybe fifteen percent higher than redline? Blunder into the unknown above that and you may become an experimental test pilot, and at a rather early point in your career.

No, as you reach the upside-down position you must push slightly *forward* on the stick, to hold the nose up, thereby increasing the sensation of hanging in your harness. As you roll through inverted (all the while maintaining that full aileron deflection), you should also be reversing the rudder deflection, since you will require rudder the *other* way to continue to hold the nose up when you are vertically banked the *other* way. (The airplane's nose will try to fall away throughout the maneuver.) As you approach level flight once more, ease off the rudder and aileron deflections. Now do you see why I said

that slow rolls are not easy? They are best practiced while you are lined up with a straight road or railroad or section line. You should also check that you are maintaining altitude throughout. You can ask a friend on the ground to tell you if the airplane looks as though it is rolling level, and not rising or falling during the maneuver.

Once proficient you can try slower and slower rolls, using less than full aileron deflection. Advanced stuff is the super-slow roll, defined as one that takes more than fifteen seconds to complete.

Are you with me so far? In truth, you should be with more than *me*. You should be with an instructor till you have mastered at least the basic aerobatic maneuvers. You *can* go out and start from scratch all by yourself, but this does involve an element of risk, should you accidentally get into something over your head—some maneuver you can't handle or see instantly how to recover from. Your airplane should, of course, be one that is specifically approved in its airworthiness certificate for aerobatics (simple trainer types usually are not), and parachutes are a fine idea. In some countries (England, for example), you may fly aerobatics without a chute, and in the United States you may do aerobatics solo without one, but if there are two of you aboard, both must by law wear chutes.

Next we'll try a hammerhead (called the "stall turn" in England). Of all simple aerobatics

Author's no-longer-new but still-agile
Jungmeister (wearing colors of 1930s German
sport-flying association) executes
what he is pleased to call a "quick, deft,
utterly reliable snap roll."

this is surely the prettiest to watch from the ground, and flown accurately it is satisfying to do. Start from an airspeed a little above cruise, with full power, in level flight. Pull the airplane firmly but smoothly into a vertical climb and hold it there, still with full power. The horizon should be bisecting each wing-tip—check this to establish that you are not leaning to one side. Making sure you are exactly vertical in the pitch axis is harder; a true vertical climb is another of those things that can really be established only after criticism from someone watching from the ground. From where you are, the wing-tip generally should look as though it is a little past vertical against the horizon. Hold it there until you run out of upward momentum, until your airspeed indicator reads zero, then kick on full rudder to cartwheel you over sideways. Just before you start coming down vertically, ease off the rudder and close the throttle. Then gently ease the stick back to recover from the dive to level flight.

If achieving a perfect hammerhead sounds easy, it isn't. Besides the difficulty of achieving and holding that exactly vertical climb at the start, picking the perfect moment to kick on rudder is a trick. This should be the moment you are motionless in space, just before you start to fall backward. But since you are in the engine's slipstream, the air will be rushing past the cockpit as though you were still going up. If you put the rudder on too soon you will wallow slowly over sideways in an untidy manner. Wait too long and you will tail-slide backward, with the controls thrashing against their stops because of the reverse airflow. Not good for the airplane's structure, this, for you can easily damage the control circuits. Your airspeed indicator may have a lag in it, and it may still be indicating some forward (upward) speed when you are in fact motionless. Again, the only way to know for sure whether you are getting your hammerheads right is to ask someone watching from the ground.

Torque comes into it, too. Torque is a perfect example of Isaac Newton's law that to every action there is an equal and opposite reaction. Your engine, grinding away at full power, is rotating the propeller in one direction, and the propeller is trying in consequence to rotate the airplane the other way. The slipstream from the propeller also follows a spiral course back around the fuselage and rudder, making the rudder more powerful in one direction than the other. Stall turns therefore always go much more easily and neatly in one direction (to the left with American engines, which rotate clockwise as seen from the cockpit) than in the other. To do a hammerhead *against* torque you usually have to close the throttle once the cartwheel has begun, and as a result it never looks so impressive.

Torque also affects rolling maneuvers, particularly the snap (in Britain, "flick") roll. A snap (the

97

maneuver we began the chapter with) is a horizontal spin done at a fair airspeed—normal cruising speed in my Bücker. With full power once again, you pull the stick firmly back till it is on the back stop, kicking on full rudder as you do so. The effect is to stall one wing (or pair of wings) on one side but not the other, which continues generating full lift. The result is a very quick roll, indeed. You recover by jabbing the stick forward to unstall the wings and kicking opposite rudder to stop the rolling rotation. (You can use the ailerons to help, but it is the rudder that really generates the rotation.) Yank and kick, in a trice, and jab and kick to recover. My Bücker snaps so fast that I actually do most snaps *against* torque, using the torque at the end to achieve a clean, sudden recovery. If you snap with torque, *it* tends to want to continue after *you* want it to stop. Snaps feel quite violent; an ordinary snap generates four g in my Bücker and six in a Pitts Special.

Variations are the double or triple snap, where you let the rotation continue for two or three turns; one and a half turns from knife-edge to knife-edge, where you first do a quarter roll with the ailerons till you are flying sideways, on edge, then initiate a snap over the top, stopping it after one and a half rotations when you are flying on edge the other way; and outside or inverted snaps, where you push the stick *forward* and stall the wing for inverted flight. You achieve an equal amount of g in an inverted snap, but, of course, it now tends to throw you out of the cockpit against your straps, rather than keep you in your seat. Something that goes well in my Bücker is to do one and a half ordinary snap rolls from level flight, recover inverted, hold it there momentarily till speed is regained, and then push one and a half outside snaps back to level flight again. The negative g and the rapid rotation make you popeyed and puce in the face by the end, but it *is* exhilarating.

All aerobatic maneuvers, however complex, are variations on the basic maneuvers I have described. For example, a hesitation roll is a slow roll wherein the rotation is momentarily stopped, then restarted at "points" (usually four or eight of them) during the roll. A rolling circle is a series of slow

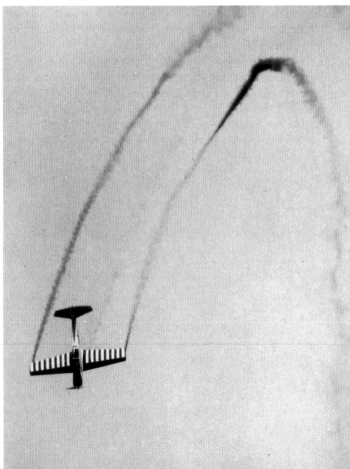

Chipmunk Special, with Aerobatic Expert Art
Scholl as pilot, rudders over at apex of hammerhead
(or stall turn) (l), then completes
maneuver with steep, accelerating dive as
wingtip smoke-flares trace the course of the stunt.

rolls done while the airplane is flying in a horizontal circle. A rolling loop combines rolls with a vertical loop. An outside loop requires you to *push* the stick instead of pulling; as you loop, your head is on the outside, and you are once more hanging hard in your straps all the way round. A vertical S is an inside and an outside loop combined. A spin is the same as a snap, but initiated at minimum airspeed and usually done for a number of turns coming straight down. (Yes, you can do an outside spin, too.) In a tail-slide you pull up, as for a stall turn, but do not apply the rudder. The airplane actually falls backward for a length or two, and you can choose which way it will flop, forward or back, by moving the elevator fully forward or fully back. (You must jam your feet hard on the rudder pedals and hold the stick firmly to prevent reverse airflow from bashing the controls violently against their limit stops and maybe damaging something.)

Just about every aerobatic maneuver in existence today had been discovered as early as the end of World War I in 1918. Every now and then somebody comes up with some trick that appears to be novel, but study of early histories of exhibition flying always reveals it to be no more than a rediscovery. The latest of these was the torque roll, used by Texan Charlie Hillard in the free-style sequence that helped to win him the World Champion title at the World Aerobatic Championships held in France in 1972. For a torque roll you pull up into a vertical climb, rolling as you go. You let it continue into a tail-slide, still rolling as you start falling back tail-first. Eventually you flop into an ordinary dive.

We all thought this was really new, until someone discovered this paragraph in *Diary of an Unknown Aviator*, the autobiography of John McGavick Grider, an American pilot who flew with the British Royal Flying Corps in 1918: "I gave my new plane [an SE5a] a work-out in the air today. It flies hands-off; I put it level just off the ground and it did 130. Then I went up high and did a spinning tail-slide. Nothing broke so I have perfect confidence in it." Wow! Remember that RFC pilots didn't even have parachutes, since the British authorities felt that furnishing them with chutes might sap their aggressiveness and will to fight!

Toward the war's end, RFC pilots were doing a comprehensive range of sophisticated aerobatic maneuvers in their off-duty flying. The Sopwith factory test pilot Harry Hawker seems to have been the leading light in this. One modern English aerobatic pilot, the late Manx Kelly, suggested: "If your life expectancy in combat was about three weeks, and to counteract the effects of castor-oil vapor from the engine lubricant you drank a lot of brandy, and you flew a tricky and maneuverable airplane like a Sopwith Camel, and you had heard that Harry Hawker had done some amazing maneuvers with it . . . then what the hell, let's try something wild!"

The other great aerobatic novelty of recent years was the lomcevak, discovered by the Czech pilot Ladislav Bezak, who was World Champion in the 1960 contest held in Czechoslovakia. *Lomcevak* is a Czech word that describes the effects on one's equilibrium of drinking too much of their slivovitz plum brandy: you tumble! And the lomcevak is a way, evidently the only way, to make an airplane tumble tail-over-nose. There are many variations on the basic idea of the lomcevak—as was brought home to me when I went to congratulate the Czech Zlin pilot after his display at the Paris Air Show in 1975. His lomcevaks were superb, I told him. He was quite offended. "Is not lomcevak," he assured me. "Is plate." What he did certainly looked like a lomcevak to me, but in truth the airplane did also resemble a plate spinning and wobbling atop a juggler's wand.

I will describe the original, or "grand," lomcevak. You push up from inverted flight into a vertical climb. You then initiate an outside snap roll in the direction of torque: full forward stick, full aileron deflection one way, full rudder the other. The airplane seems to come to the hover, with the nose first tracking wildly around the horizon, then pitching down and underneath and up and over the top. This can continue through several complete rotations. What seems to happen is that gyroscopic precession takes over and changes the axis of rotation till you tumble end-over-end. The spinning propeller and crankshaft form a kind of gyro, and if you apply a deflecting force to a gyro it tends to move in a direction 90 degrees on from the direction of the force. The "deflecting force" here is your control in-

99

put. The maneuver ends when the airplane simply runs out of rotational inertia, and becomes virtually motionless. It may at this moment be pointing in any direction, up, down, sideways, but it quickly falls into a dive, regains forward airspeed, and you can recover to ordinary level flight.

A lomcevak is a hairy experience for the pilot. You would hardly believe the amount of negative g, the speed of the rotation, or the aerodynamic buffet that shakes you. Nor is it good for the airplane. Many of the air-show pilots who used to demonstrate lomcevaks as a matter of course have given them up because of split fabric on the wings or damaged propellers, engine cranks, or bearers. Once again, the lomcevak cannot even be called truly "new"; rather, it is a rediscovery. Let us go back to 1918 again, and John Grider: "I was next and I put my nose down to about two hundred after I did my full roll, and as soon as I started up for my full zoom I kicked on full right rudder and pulled the stick back into the right-hand corner. I didn't know what I was doing but I sure did it. I whirled around with my nose down and ended up stalled upside-down. The motor stopped and I did just get into the field with a dead prop. It was Thompson's turn next. Mac said what I did was an upward spin followed

by an outside spin, whatever that is. I told Thompson how I did it and he went up and started into it with terrific speed. The propeller shaft broke and his prop flew off just nicking the leading edge of the wing. He got into the field alright. That ended the afternoon performance. Mac and Cal can certainly fly." That "upward spin" was well on the way to being a lomcevak.

There is a notation for aerobatics, a set of symbols used for drawing out a sequence of maneuvers on paper, much as a violin solo can be drawn out on a musical stave in a form all musicians can understand. It is called *Sistema Aerocriptografico Aresti*, after its inventor, José L. Aresti, a Spanish count who was also one of the great exponents of air-show flying in the Bücker Jungmeister. His "system" is the beautifully thought-out culmination of years of work and development, and it has been universally adopted for contest use by pilots of every country. Not only does the Aresti system make it possible for every maneuver to be drawn, it also assigns each one a "difficulty coefficient," or K rating, so that contest judges can standardize their scoring of aerobatic performances. Thus, if a maneuver is rated K-20, a pilot who does that maneuver half as well (or 5 out of 10) as the judges think it could be done scores

A lomcevak. This Czech word, meaning the imbalance that results from imbibing slivovitz, aptly describes the tail-over-nose tumbling performed in this spectacular aerobatic stunt. Several end-over-end rotations subject pilot to high negative g and aerodynamic buffeting, and subject airplane to great stress.

100 points (20 times 5)—the same as one who does a much easier K-10 maneuver perfectly (10 out of 10). Before the Aresti system, contest judges could only rate a pilot's total performance, after he had finished flying. Now every single maneuver can be appraised as it is completed. (Usually a contest has several judges, each watching the flying and calling out his or her appraisal to assistants with a pen and paper.) Aresti's notation system also enabled set programs to be drawn up and published for pilots to practice in advance. Big national and international contests now have four "programs": the "known compulsory," published well in advance of the contest, which each competitor may practice to his or her satisfaction; the "unknown compulsory," drawn up by an international contest jury the day before it is to be flown, which every contestant must fly

without prior practice; a free-style routine, where each competitor prepares his or her own sequence, an Aresti drawing of which must be filed with the judges in advance; and an "unlimited free," in which the contestants can do anything they wish and are judged on the artistry of the total performance rather than on individual maneuvers—as though the Aresti system did not exist, in fact.

The elaboration of contest rules made possible by Aresti has undoubtedly improved the standard of aerobatic flying world-wide, and has made judging much more precise. It has also made contests—at least the first half of them, where every pilot flies the same routine—a bore for spectators and a wearisome chore for contestants and officials. It was to liven up the proceedings that Group Four (the unlimited freestyle) was first introduced at the 1972

Two of world's finest aerobatic planes are
Czechoslovakia's clean-limbed Zlin 526 Akrobat
(below) and Soviet Union's Yak-18 (far r).
Zlin can do more than Jungmeister,
though not so easily. Powerful, steady
Yak is a consistent winner. Right:
Pilot's intended sequence of maneuvers
is taped on cockpit instrument panel.

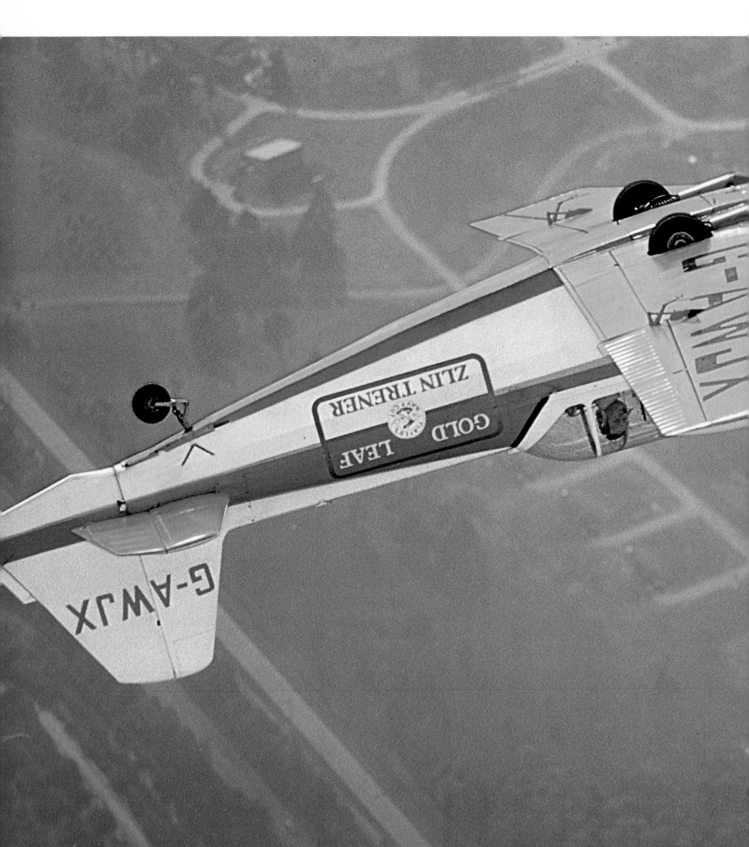

World Contest in France. And if Aresti has created an ever-widening gulf between contest aerobatic flying and the type of flying done purely for entertainment at air shows, then this new Group Four is an attempt to bring them closer together.

The Aresti system has its weaknesses. It has always seemed to me that it assigns too low a difficulty coefficient to complex snap maneuvers and too high a K to outside looping maneuvers—perhaps because the Jungmeister, José Aresti's own mount, snaps well but performs badly when flying "outside." It attaches too much importance to the lines and angles on which pilots fly their maneuvers, producing somewhat stilted performances, rather than the more gracefully linked continuous routines we saw before Aresti. But its universal adoption by every country that competes internationally shows the strength of the Aresti system. I like Aresti's dedication, printed in the large and complicated book with which he launched it: "To all pilots fond of aviation that with their ardor, enthusiasm and great sportiveness did contribute to the graphical representation and actual perfection and development of Artistic Flight all over the World." Olé!

The Jungmeister, though a joy to fly and possibly, from the handling point of view, as superb an aerobatic airplane as any ever built, is no longer competitive for contest flying. It evolved as a smaller, single-seat, lightweight derivative of the Luftwaffe's basic two-seat training airplane in the mid-

thirties. Jungmeisters won most of the rather amateurish (by modern standards) contests of the 1930s. And it was, I think, sponsored by Hitler's Germany largely for exhibition flying as part of the Nazi cult of the Aryan superman. While basic aerobatics were taught to the hordes of pilots trained by the military of every nation in World War II, there were no contests or air displays during that war. As a sport and pastime aerobatics revived slowly in the postwar years. At first only ex-military trainer aircraft of doubtful suitability were available: Stearman and Great Lakes biplanes, Piper Cub and Taylorcraft monoplanes in the United States; Stampes, Tiger Moths, Jungmanns, and Jungmeisters in Europe. The first signs of real progress in the evolution of new aircraft types came from the most unlikely of places, Czechoslovakia, then lately fallen under the Communist yoke. Starting with a wooden German Bücker design, the Czech state aircraft factory re-engineered the plane in metal and installed a Czech engine, thereby creating the Zlin aerobatic monoplane. The Czechs played host to the first-ever World Contest in 1960, largely I think to display to the world of sport flying just what they had wrought —a convincing display. For a decade their Zlins were unbeatable and were bought by foreign pilots in the West, as well as by the state aero clubs of the Communist countries. The Russians evolved a series of aerobatic special versions of their own Yak-18 basic trainer at the same time. It is significant that the

Left: Dangerous, rarely performed back-to-back loop. At top of maneuver, in extremely close quarters, inverted (striped) plane is making a conventional inside loop, starred plane is heading into more difficult outside loop.
Right: Tail-slide is falling back, tail-first, from vertical climb until level, then diving forward.

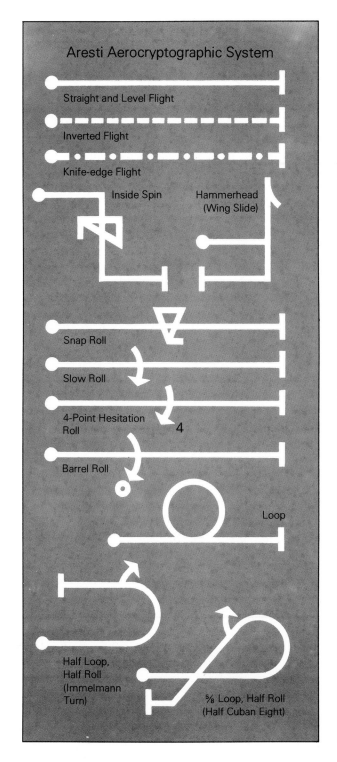

Aresti Aerocryptographic System

Straight and Level Flight

Inverted Flight

Knife-edge Flight

Inside Spin

Hammerhead
(Wing Slide)

Snap Roll

Slow Roll

4-Point Hesitation
Roll

4

Barrel Roll

Loop

Half Loop,
Half Roll
(Immelmann
Turn)

⅝ Loop, Half Roll
(Half Cuban Eight)

Only full-time civilian aerobatic
team, sponsored by Rothmans,
flies Pitts two-seaters, very
proficient and responsive aircraft.
Down and up pictures
(l & r) were taken during unison loop.
Right: Fisheye-lens view of
''double mirror'' from level-flying rear plane.
Below & bottom: Box formations
with one plane inverted.

107

first five world contests were all held in (and won by) countries with totalitarian governments, for it seemed only countries with centrally controlled economies could afford, by simple allocation of resources, to fund both the contest and the effort required to win one.

American sport pilots, used to the unquestioned superiority of everything American, found this galling. And while the United States did eventually, after a long struggle, learn to master the Aresti-system basis of the contests to the point where it won all three titles at the 1972 contest, it has never managed to overcome the economic problems involved in playing host to a world contest. (How would you transport all the European contestants' airplanes to the States, for a start? The U.S. team has used USAF C-130, C-141, and C5A military freighters to get their own aircraft to European contests, but among the European nations only the Russians have flying freighters big enough to transport their own machines across the Atlantic.)

In the United States aerobatic lightplanes have all come from the home-built airplane movement, the market for sport planes apparently too small to interest the big manufacturers. The most astoundingly successful design has been the Pitts Special, now in series manufacture but originally a home-built design that grew from Curtis Pitts's ambitions in the 1930s to be an aerobatic champion. As a pilot he never made it, but he did achieve the crowning satisfaction of seeing American pilots win all the prizes at the 1972 World Contest flying his airplanes. Equally remarkable was that they were biplanes, for monoplanes, mostly Zlins and Yaks, had won every contest for the previous decade. Monoplanes are aerodynamically cleaner, with lower drag, and generally perform better in the upward-rolling maneuvers that now rate so highly in contests. Biplanes, because of the cross-bracing that the double wing structure allows, can be built very strong and light, and Curtis Pitts showed that a small biplane with a big engine could achieve a performance comparable to a monoplane's. For the last few years the Pitts has been supreme in the United States, to the point where hardly any serious contestants ever competed in anything else. Then Leo Loud-

enslager won the 1975 U.S. National Contest in a home-built Stevens Akro monoplane—so perhaps the trend will reverse. Certainly the Czechs, whose Zlins had been eclipsed by the tiny Pittses, have come back fighting with an entirely new model, the Z50L, which has an American engine. The other important new design of recent years was the German-Swiss Akrostar, a high-powered monoplane with a weird control system that involved a kind of variable-camber wing—trailing-edge flaps that moved up or down in opposition to the stabilizer, and differentially with the ailerons. The Akrostar could do things impossible in any other airplane. I once watched one pull up into a vertical climb, then roll 90 degrees left, 90 right, 90 left, and 90 right before running out of upward momentum. But it was so expensive to build, and so unstable and tricky to fly, that production ceased after a dozen or so had been built.

At one stage the Russians seemed to be trying to promote civilian jet aerobatics; but even a small jet is beyond the financial means of individual sport pilots. The military have their jet aerobatic teams—the USAF Thunderbirds, U.S. Navy Blue Angels, RAF Red Arrows, Italian Frecci Tricolori, French Patrouille de France, Belgian Slivvers, and so on—but you'll notice that everybody keeps quiet about what the teams cost. The justification is that they attract new recruits to the services. The more powerful reason, perhaps, is military prestige.

Aerobatics are the song-and-dance routines of flying, the stepping forward to the footlights, the beginning of your speech. It is showing off, pure exhibitionism. An air-show aerobatic routine is nothing more or less than a huge boast of your ability as a pilot, a declaration that you have passed beyond any fear, an assertion that what you can do in a plane (make it sit up and beg) is worth watching, and a demand that people watch. There is also much of the appeal of a fairground roller-coaster ride in it, the simple ballet dancer's exhilaration in whirling yourself around and turning the world upside-down. Then there is the camaraderie of contests, matching your mettle against others, then learning what style of men and women they may be. Aerobatic flight is certainly *my* favorite corner of the flier's world.

108

Eleven Fouga Magister jets of the French Air Force aerobatic team, Patrouille de France, fly low, close, and fast in precision maneuver at air show. Wingtip-to-wingtip flying at speed is difficult and dangerous. And military flying teams are extravagantly expensive.

Destinations

Flying, as befits a mode of travel that has made it possible to get to the other side of the world in just a day and a half, is a universal activity. The pattern of a journey varies little from country to country—especially now that English has been adopted as the universal language of aviation. It is generally well spoken, too, except in France, where they still think French should have been chosen; in some regions like the Caribbean, where it is spoken with such a strong accent you may hardly recognize it is English; and in a few countries, Russia for example, where because of political considerations there are so few foreign movements they have no English yet.

Many aviation terms do differ from land to land. An airport traffic pattern is called the "circuit" in Britain, and gas in the U.S. becomes "petrol" in Britain, but *essence* in France (where *pétrole* means the raw stuff, straight from the oil well), and *benzina* or something similar in many other languages. Most aviation terms, though, are surprisingly similar even in strange tongues. I once saw a huge T-handle projecting from the instrument panel of a Russian airplane, with Cyrillic characters on it proclaiming its function, and thereby learned that the Russian for "starter" is *starter*. (But *starter* in French is the choke in an automobile.) American aviation terms are tending to oust native words in many parts of the world; even British pilots now lower the gear, instead of the "undercarriage."

Traffic patterns vary, too, if you cannot get a straight-in approach and must fly in visual pattern. In the U.S. you enter at 45 degrees to the downwind leg; in Britain you first come up over the field several hundred feet above pattern, or "circuit," height, and head off upwind before turning across—then downwind and descending to pattern altitude. Over Britain you may only follow an airway if you are under instrument rules—quite unlike the United States—and over France and some other European countries you may not fly by visual rules at night. "Homing" letdowns, with guidance provided by the tower controller taking bearings on your voice transmissions, are still available at some smaller European airports, but unheard of in the U.S. Avionics requirements vary: you need an ELT emergency transmitter for all flights in the U.S. but for none in Europe; and DME and altitude-reporting transponders are required for IFR in some regions but not others.

The variety of ground radio stations available for navigation surprised me the first time I flew in the U.S. There seemed to be modern VORTACs everywhere, and instrument-landing systems at the smallest airports. Simple nondirectional radio beacons are the only en-route (and even the only letdown and approach) aids in some less-developed regions. Simply filing a flight plan is considered sufficient notice for most international flights, but I remember one country, Cuba, that required you to send a reply-paid telegram request twenty-four hours in advance, and some Arab and African states require you to apply well in advance through diplomatic channels. Over Albania you may not fly at all. And many countries—even a few quite liberal ones like Sweden and West Germany—do not permit you to take photographs while flying over them.

The fees mulcted for airways and airport services and immigration and customs clearance, and the unhurried pace at which these are provided, may occasionally leave the aviator fuming. In America they come out in a yellow airport automobile to hit you for a landing fee where one is charged; elsewhere you are usually expected to "report to the tower" and seek out the tormentors of your wallet. All countries reserve some air space for their military, and those with seacoasts often have buffer zones and require you to identify yourself by radio to some radar-watcher before you enter them. There are also a few pinprick areas of forbidden air space. You may not fly over the White House in case you have evil designs, or over established bird sanctuaries at low altitude, or over certain research establishments where experiments are conducted with lasers or radars of a potency sufficient to damage you or your machine.

Weather information nowadays is generally provided in coded teleprinter messages. These codes may be unfamiliar to you, and you may need to find an official to translate them for you. Weather is not even always called weather. It is "met" (for "meteorology") in much of the world, and it can be all the more confusing when the "met man" across the desk

Preceding pages:
Unobstructed view of splendid
Kansas sunset from
privileged seat in Piper Cherokee,
a modern four-seater.

from you (I am thinking particularly of France) speaks no English. You can usually puzzle out the information you need, but it can take considerable time. Weather is . . . well, weather, wherever you find it, and universal enough, but it can still present surprises. Until you have experienced it, you may not appreciate just how smog in the Los Angeles basin or fog in the valleys of Switzerland can reduce an otherwise lovely day to really instrument conditions, or how strong winds over mountains can generate fierce turbulence and make it impossible to hold altitude in the up- and downdrafts at lower levels. Having trained in England, I had no idea just how unlimited unlimited visibility could be until I flew across the American West and found I could clearly pick out mountains that the map showed to be seventy miles away. Nor, for that matter, did I realize how suddenly a midwestern summer thunderstorm could reduce ceiling and visibility to zero. Snow showers, similarly, can quite suddenly reduce generous visibility to nothing—I know one pilot, trained in the Bahamas, who came to grief flying over France in midwinter. But the well-read aviator has a nodding acquaintance with such unfamiliar hazards from his textbooks.

One embarks on a flight to new lands, new territory, with a special relish. One such for me was my first trip down to the Caribbean in a lightplane —a journey that took full advantage of the airplane's unique ability to turn winter into summer in a few hours. The four of us had begun by assembling at La Guardia in the bitter December dawn, with the frost searing our faces. Our Cessna 320 Skyknight twin, lodged in its parking place, seemed a spidery silhouette against the low sun, whose milky morning light gleamed on the greasy concrete apron as on a cobwebbed autumn meadow. We signed for the gas, threw our bags aboard, climbed in and started up, and were soon cleared by Ground to join a line of waiting airplanes. There are morning rush hours in aviation, too. Behind us in the line was an angry-sounding Gulfstream I executive turboprop, the howl of its turbines adding to the muted drumming of our own idling piston engines. As the DC-9 airliner at the head of the line was cleared for takeoff we heard the sudden roar of its engines—a mad ca-

cophony. A Mohawk airplane hurried past us and turned into the run-up area; the captain's storm window opened to emit a shirt-sleeved arm holding a container of coffee, whose unwanted dregs spattered to the concrete below. I have always had this fantasy that Mohawk airplanes ought all to be flown by Indians.

Like divers on a springboard we edged up slowly, waiting our turn to leap into the air. It came at last. As we turned south on course Manhattan passed below, still lying dreaming in its smoky morning halitosis. A light frosting of snow covered New Jersey, as though a giant had been busy with a colossal sugar sprinkler. The sun gleamed bright and pale on the wing, and we seemed to hang motionless as a fly in amber.

To refuel we descended at Charleston—a kind of landing place common enough in Europe but rare in the States, a jointly operated civil-military field. Before we could depart, we had to wait a moment for the shock waves of an afterburning Air Force jet to subside. South of Charleston the landscape, seen from above, appeared to be no more than three soggy inches above sea level. Bangladesh has the same look, and so does the Amazon. We landed next at Fort Lauderdale, dismounting stupidly into the eighty-degree afternoon still in our overcoats and northern pallor. It was my first time in Florida, and from the taxi I thought it disappointing: shacks done up as night clubs lining the highway, and billboards advertising nothing one could ever want. Florida, I thought, is only New Jersey with palm trees—though I liked it better later. At dusk I joined a horde of chattering matrons in the hotel bar, a Stygian cavern lit only by the submarine lighting of the swimming pool, which formed one wall, and the mauve flickerings of the inevitable color television set, on which I watched a Korean choir in full national dress sing "Coming Through the Rye"—not something you can see everywhere, every day.

I woke the next morning to find the air still smelling deliciously of summer, of green things growing. I had half expected it wouldn't, that the previous day's unaccustomed midwinter warmth had somehow been contrived just for our arrival. New York had smelled only of frost and dirty air.

113

Lauderdale's green airport was speckled with motionless white egrets. Were they real? Not till one moved was I sure. We began our interminable dealings with the customs people, who scratched at sheaves of useless forms documenting the aircraft and our cameras. Nation shall speak unto nation, but it shall be in triplicate.

Radar had us "contact" almost before the landing gear had banged up for its long sleep till Eleuthera in the Bahamas. We climbed confidently into rain clouds long as an old lady's skirts. Between showers we were granted glimpses of that famous island sea, inky blue where the water was deep, sky-pale and bright over the bleached coral shallows. There was vivid sunlight on Bimini, showing a long runway and a tantalizing swimming pool.

Then the cumulus thickened to gray stratus and rain as we penetrated a stationary front; but it cleared by the time we reached Eleuthera. I studied the chart. "Rock Sound International Airport," it an-

nounced grandly, leading me to imagine something more imposing than a single low-powered radio beacon, a wet and wavy runway, and a rough area of crushed coral to park on in front of the low wooden bungalow that served as a terminal building. Here, enjoying the trade wind rustling the palms and the antics of a hummingbird shyly plundering the airport's flowers, we waited an hour for a reluctant customs officer to come out from town. Our taxi didn't mind waiting, either.

Eleuthera is flat, covered with a mean and hungry scrub, but as we neared the Rock Sound Club we saw tall trees and flowering shrubs. From the club's office we walked down a flowery path to discover a superb pool framed with palms and flowers, and a cool restaurant, and other jungly paths leading to our quarters. Elsewhere were cottages on a magical beach that also belonged to the club; and much farther on, down a road that ran ruler-straight but bounded up and down over the low hills like a

switchback, a crumbling Bahamian village, a frightening slum of rotting wood shacks littered with those two world-wide signs of poverty—goats and wrecked automobiles. We returned to the club to dine in its richer atmosphere of parquet floors, dark curtains, candles in jugs and plants in pots, and reassuringly obsequious waiters. Then we sat out in the tropic night and talked to the club's manager, who told us that he employed four Swiss chefs, could accommodate one hundred guests, and that about a tenth of his two thousand visitors per season came in private planes, not on airliners. He had a poor opinion of us flying "island hoppers," complaining that we are vague about reservations, and inclined to call him suddenly on his radio and haggle about prices, while we circled around the airport awaiting an answer. But to me, at that moment, it seemed the only way to travel.

Next morning we discovered the principal drawback of island hopping: you often have no communication with anywhere and can get no weather information. We took the air simultaneously with the arrival of a monster rain shower, and tried to raise Nassau Radio from under it at 500 feet. Silence. After thirty minutes we forced our way up visually through a hole in the heavy, mountainous clouds, and at 5,000 feet Nassau could finally hear us. Two lazy controllers tried at first to refer us to each other's frequency, but we persisted, and they allowed us to file instruments to South Caicos. We maintained 5,000 feet, sunshine and showers, cruising at better than 200 mph. By San Salvador the weather picked up and things began to look more cheerful (as they did at this point for Christopher Columbus), and thereafter we cruised in the clearest blue, lightly speckled with white cumulus, with convoys of coral cays below. Flying is primitive in these parts: no Omnis or DMEs or radar—just an occasional old-fashioned NDB beacon. But the Caribbean, like all oceans, is a friendly place aloft, and there is usually some airline jet way up at Flight Level 370 happy to chip in and relay your position report from his superior altitude. Thank *you*, sir.

South Caicos proved to be a distant outpost of the British raj, all salt and sunshine, with a fierce cross wind (like most island airports, I think) and an additional and unusual hazard: heaps of donkey droppings down the runway centerline.

Onward. This time the NASA moon-watchers on Grand Turk Island filed our flight plan for us. The weather at San Juan? Here a friendly Clipper captain interrupted to say that from what he had just seen of it, passing at 31,000 feet (9,450 m), it looked clear as a bell. As we neared the little Caribbean outpost of the United States, the ether began to show signs of civilization. Omni needles began to dance, the DME gave a silent *click!*, awoke from its daylong sleep, and began rolling its digital eyes at us, and our transponder began a winking flirtation with San Juan Center. Soon we were in radar contact and gifted with traffic advisories just as at home. We landed at St. Thomas in the American Virgin Islands.

Here, for all the jungly mountains that surrounded the airport, we were very much back in the U.S. as we parked amid rows of familiar N-numbered shapes. The drive into town, the beautifully named Charlotte Amalie, revealed, besides the inevitable Caribbean shantytowns, such signs of civilization as a little industry, a traffic jam, and commercials on the radio.

We arrived at our hotel as the sun went down behind the cloud bank covering Puerto Rico beyond the bay. It was one of those fabulous tropical sunsets, presented as though for our benefit. Waves of brilliant color, orange, crimson, lemon, and mauve, slowly flooded across the sky in turn. We watched in silent astonishment. We were still drunk with color at midnight, when we ended our evening at McCleverty's Fallen Angel (Showcase of the Stars), a night club with beautiful wrought iron chairs, lit by shining globes of red, green, blue, and purple hanging over the bar. And there are, I had by then discovered, more things you can do with rum than just mix it with Coke.

The tidy Danes ruled St. Thomas until 1917, and as I wandered on foot the next morning I decided they had left the United States a legacy of timeless beauty in Charlotte Amalie.

By noon we were winding up our engines to takeoff power at the runway's start. You depart here over a famous mountain that sheds some horrendous bumps for your climb-out. IFR again, we were soon plunging in and out of fat tropical showers, some of which seemed to me to have hollow centers and a spiral form, as though they nursed ambitions to grow into hammering hurricanes. We had no steady view of the passing islands, but occasional magical glimpses down through the broken clouds of Tortola, Virgin Gorda, St. Martin, Saba rising tiny and cone-shaped from the milky sea, St. Christopher, Montserrat, Antigua, and, finally, our destination—the French island of Guadeloupe. Two islands really, facing each other across a mangrove swamp, one a tall volcanic mountain ridge, the other flat coral, low as a *crépe suzette*. Once landed, we found ourselves miraculously in the real France, for Guadeloupe is no colony, but a *departement* of metropolitan France. Even the airport had all the appurtenances of French life: tricolors, unintelligible loudspeaker announcements, uniformed officials, Gauloises and garlic, inquisitive flop-eared dogs, and some lunatic buzzing the tower in a police helicopter loaded with giggling girls. Going into town we passed the usual shantytowns, called *bidonvilles*, jerrycan-towns, by the French, mixed up with modern apartment blocks straight from the outskirts of Paris or Lyon. A sudden downpour fell upon us and the car filled with steam, which cleared to reveal our hotel, La Caravelle.

In this faraway landscape of tropical foliage and neglect we were not prepared for La Caravelle. Imagine if you can the TWA building from JFK spirited away and set down here at the water's edge as a hotel. The roof was like a great seashell balanced on spindly curves of concrete, and the hotel's public rooms had no walls and no windows, but were open on every side to the trade wind gently sighing through the palm trees, around the pillars, playing with the papers on the clerks' desks. The dining area surrounded a huge cold buffet of giant lobsters and shining mayonnaises tinted green or pink. By it stood four chefs happily explaining their complex handiwork and basking in the glory of their achievement. Beyond the chairs and tables, parading on the hot sand, was a half-circle of grackles, beautiful clowns, gossiping and chirruping impatiently while they awaited our crumbs and leavings. Beyond them

the pool, and a concrete ramp climbing to an upper floor, its railing slowly surrendering to the bear hug of a mauve morning-glory. Beyond that, palms and sea-grape trees drooping in the heat down to the sand; past them the ice-blue glittering hot sea. It was like one of those half-forgotten childhood memories of a moment and a place so superb you wished you could stay forever.

Our departure two days later began with a visit to the "Meteo," a stoical Frenchman whose idea of a weather briefing was to roll the stub of his Gauloise to the corner of his mouth, wave his arms expansively, point to the mingled sunshine and cumulus outside, and say, simply, *"Comme ici, m'sieur."* Pressed for greater detail, he told us the tops of the cumulus would be at 6,500 feet (1,980 m) in which he was no more than 3,500 feet wrong. Would you believe 10,000 feet (3,048 m)? Our climb-out from Guadeloupe was interrupted by a cough from one engine—water in the tanks? Apparently, for we jiggled with the mixtures a bit and all was again perfect. It was the only less-than-perfect behavior from our Skyknight on the whole trip.

Back at San Juan we faced a new Inquisition from the U.S. Customs officer, even though we explained we were in transit and had stopped only to take on fuel. We went on toward Jamaica, skirting the northern shore of Puerto Rico and the southern coast of Haiti and the Dominican Republic. At Flight Level 80 we were skimming a solid sea of strato-cu, our engines warbling pleasantly at 2,200 rpm and 27 inches. The eastern part of the Dominican Republic appeared in a rift in the clouds, forested, flat, empty, and lovely. The Caucedo controller was almost unintelligibly Spanish, and his compatriot in Santo Domingo sounded as though he had a mouth full of banana. Haiti made not even a pretense of exercising control over its Flight Information Region, and studying the chart I saw a warning not to fly over the presidential palace in Port-au-Prince, or you might be summarily fired on. The Kingston beacon then came in on the ADF, right on the nose, and an hour later the controller there answered our call and sent us up to ten point five to be clear of the coffee-covered Blue Mountains. Jamaica, when it hove in view, wore a heavy canopy

117

Officials and staff pose
obligingly at Jamaica's Montego
Bay airport. Right: Loading
Skywagon for takeoff
to Grand Cayman Island.

Even from 500 ft (152 m),
poverty of this Russian village
near Minsk is evident.

of cumulus that proved, when we were cleared to descend into it, to contain some fierce bumps. We landed at Montego, stretched and yawned after four hours aloft, and drank a fine free rum drink courtesy of the Tourist Board. Said the fuel man, examining our credit card after he had put 102 gallons into our 140-gallon tanks: "Shell *domestic*, that's really for use only in the United States, mon . . . but I dare say we can use our discretion. You can't go through life without discretion. It's the better part of valor," and we didn't disagree.

Jamaica, I learned with astonishment, has some forty airstrips and landing grounds, and I visited half a dozen of them later in a Skymaster belonging to Jamaica Air Taxi, a little company operated by an American expatriate. Precarious strips, they were alongside bauxite mines up in the mountains or green meadows in banana plantations down by the coast. Timson Pen, a short and bouncy coral runway, appeared to be in the middle of downtown Kingston. The giant jets use Kingston International, whose airport restaurant serves the best cup of coffee on earth. Blue Mountain coffee, mon.

In a single-engined Cessna we later made a foray across two hundred nautical miles of ocean to visit Grand Cayman, a flat, dull coral strip that once was home to a million turtles (now mostly eaten) and is distinguished by the odd and enviable quirk of levying no taxes. It seems in consequence to have as many banks as it once had turtles.

In the heat of our last day in Jamaica we stumbled along to Doctor's Cave, a tiny sheltered beach where we dozed happily under the sea-grape trees till we were roasted pink, then rented goggles and flipper feet and went snorkeling in the slow green sea, floating over the rich coral, a realm more strange and fantastic than the backside of the moon.

A mild adventure was on the flight plan for our next leg. We were to fly across Cuba. But we had telegraphed for permission and there were no problems. The Camaguey controller was quite unexcited by our passage, though he did ask our home base. The spelling of "Wichita" seemed to flummox him. We did pass a Cuban airliner a thousand feet below, a Russian-built Ilyushin. An hour more took us to Nassau, whose airport was a busy jumble of general aviation aircraft, with long rows of parked business jets. We went on to Fort Lauderdale to return the life jackets we had rented, then headed grimly north, back into winter. It was 81 degrees when we left the Fort, 25 and snowing—56 degrees colder—when we landed at Raleigh-Durham three hours later, feeling stupid in our suntans and open shirts. New York no better—but then it never is.

That marvelous journey was in 1966, and though I have been back to the Islands since, that first time was the best and maybe the most memorable aerial journey I ever made. A true spendthrift would not have counted the cost of such a millionaire's vacation, but out of curiosity I did. The Skyknight's owners reckoned its operating costs at $32 an hour, depreciation at about $58 an hour. We flew thirty-one hours for, say, $2,800. Such a trip by airlines, if you could have found flights to match ours in a nine-day schedule, would have cost the four of us, as I later discovered, $1,800 tourist or $2,500 first class—and in the Skyknight we were undoubtedly flying first class. On hotels, meals, planter's punches, and the like, I spent $330 over the nine days. (I suppose now you would have to double or triple those prices.) Not out of the question for the trip of a lifetime.

The Skyknight, with its 200-knot cruise and easy four-hour range, had just the pace to match those island distances. And one does feel more at peace over water with two engines. I've never *had* an engine problem in a light aircraft, but somehow all those sharks that bask so boldly in the gin-clear shallows around the Bahamas have a dreadfully hungry look when seen from behind just one fan in a Cherokee or Cessna 172.

In the world of aviation you find yourself powerfully drawn to places you would never otherwise dream of visiting. I've never found anywhere so remote it couldn't be reached by other means of transportation, but I've been to one that came close: the only other access, I was told, was by two weeks on muleback. This was Gilgit, a settlement in a Himalayan valley in northern India, nearly equidistant from the borders of Pakistan, Afghanistan, and China, that seems little changed since the days of Alexander the Great. We flew there for lunch in just

over an hour in an old Pakistan International Airlines DC-3, from Rawalpindi on the hot Punjab plain. We followed the Indus Valley, at first over low foothills sunk in the dust and heat haze, climbing hard all the time, till at 10,000 feet we were finally in clear air. The mountains grew ever higher, their blanket of fir trees giving way to snowdrifts and ice. At 10,000 feet we were a mile above the valley floor, itself a mile above sea level. One colossal pyramid of rock, the mountain called Nanga Parbat, the Naked Lady, reared nearly vertically to 26,000 feet (7,945 m)—three miles above us. A DC-3 is not a small airplane, but ours was stunningly dwarfed by that landscape.

For a while I sat in the jump seat between the two pilots, neat in their snow-white shirts and trousers. Captain Aziz even had a hand towel in his lap. For my benefit he pointed out anything he thought of interest. Once it was a couple of passing eagles, once the incoming flight, a DC-3 returning from Gilgit and resembling against the immense backdrop of Nanga Parbat nothing so much as a tiny fly inching across a whitewashed wall. More thoughtfully, he pointed out some wreckage on a rock face. It looked like crumpled silver paper. It had been a DC-3 that found its way closed by cloud, and turned back only to find the valley behind also filled with cloud. It

had climbed while circling in the clouds till it could climb no higher, then set off blindly southward, hoping to be lucky enough to miss the mountain peaks. It was unlucky.

The landscape of Gilgit itself was a high mountain desert, much like those of Iran or even Nevada. The airfield was at the bottom of a fantastic valley —no wind, zinnias growing wild around the control tower, and the sound of rushing streams everywhere. In a Land Rover we drove for a distance up the valley toward Hunza, supposedly the site of legendary Shangri-la, till the mule track dipped and turned and vanished between two gigantic boulders too close for our vehicle to pass between. We returned to the little town for a lunch of mild Kashmiri chicken curry, which was interrupted by sudden word from the aerodrome that the weather was worsening and we must return to Rawalpindi at once or risk being stuck in Gilgit for weeks. Why not? I thought, but soon we were airborne again and climbing hard.

This route was VFR only; no radio aids to navigation; no voice communications, since the mountains cut out line-of-sight VHF—just wireless telegraphy in Morse code; essentially valley-flying all the way, since most of the mountains towered far higher than a loaded DC-3 could climb. And the weather,

121

Right: Rural Poland.
Below: Hard-working, low-slow
An-2 is a large, fat Russian biplane used
for surveying, crop-dusting,
dropping sports-club parachutists.

like mountain weather anywhere, could change with amazing speed. Indeed, the fine weather and lazy winds of our trip out in the morning had now given way to thickening cloud formations and 30 or 40 knots of wind. Our return was not by the valley, but along the "high route," the direct track almost straight across all but the highest ranges. At 16,000 feet (4,877 m), as high as the DC-3 could get, each of us was breathing into his own personal oxygen bottle, for the airplane had neither pressurization nor an integral oxygen system. Most of the handful of passengers were Kashmiri girls in black robes speckled with colored stitching and round bits of mirror, dresses now popular in the West, though worn without the purdah veils of these Moslems. We were all strapped down as tight as we could bear, for the turbulence was awesome. We would shoot up perhaps several hundred feet in an updraft, then crash back down again in a couple of seconds. You had to clutch your oxygen bottle to keep the g forces from taking it up to the ceiling.

I began to be truly nervous, for I had never experienced turbulence like this, and I was afraid the old DC-3 might break up in the next hideous vortex. But the crew seemed unperturbed. At one point we flew through a plume stretched by the wind in the lee of some jagged peak, and I swear you could hear the ice rattling against the airplane. We turned into the Babusar Pass, still at 16,000 feet, and two ice-blue lakes passed by 1,500 feet below us. Then slowly the fantastic landscape began to diminish and the turbulence with it, and I began to breathe again. I would not make that trip again, yet Pakistan International Airlines puts its newest pilots on the route to gain experience—and they have established an excellent safety record on it. Now they operate their Kashmir service with Fokker F-27s, pressurized turbine aircraft that can climb above even Nanga Parbat, a major safety improvement.

Few western pilots have flown to Communist countries, yet lately there have appeared a few chinks in the Iron Curtain through which one might creep. My chance came with the 1966 World Aerobatic Contest in Moscow, which I attended as an observer with the British team. I began the journey in a Beech Baron twin that was accompanying the aero-

122

batic machines, flying first from London to Born-holm, a Danish island in the Baltic. Then to two small grass gliding fields in Poland—Lisie Katie and Datke—chosen as refueling points for the contest airplanes. The Poles were courteous and kind, but their poverty could not be hidden. You could fly above a tree-lined road for five miles without seeing an automobile—just horse-drawn farm carts. And in the restaurant washroom there was just one piece of soap, lent to you by a crone on your way in and demanded back before she would let you leave.

Borders of Communist countries are real barriers. That between the Lithuanian SSR and Poland was a high wire fence, a wide plowed strip, perhaps mined, and the rutted tracks of patrolling vehicles. We called Vilnius in English for permission to land and received in reply only a torrent of Russian. We circled the airfield, a busy airport with civil jets waiting, still hearing nothing over the radio that we could understand, and finally landed without permission. We could find no kind of formal reception party, but were immediately surrounded by a crowd of Russians, mostly in various kinds of Aeroflot uniforms. None spoke a word of any recognizable western language. No customs, immigration, interpreter, or anybody who knew what to do with us. We waited, bemused. They were obviously fascinated with our Baron, and one managed to convey that he wanted to know its speed. We wrote the answer, in kilometers per hour, in the dust, and he was impressed.

After an hour without anyone appearing to arrest or welcome us, we walked into the terminal and managed to find a nervous Aeroflot girl who located the officials we needed. It seems we were a day early for those assigned to welcome us. In Russia is forbidden to arrive early.

Later I got to talking to a group of Soviet citizens through one who had some English, and we fell to comparing living costs. When I translated into rubles how much I earned, they were astonished and downcast. When I told them how much I paid in tax and rent, and what my clothes had cost, and what the Baron had cost (though it wasn't mine), they grew more cheerful. Comparatively their salaries were tiny, but so were their living costs. That I was

123

Air pilgrim's view
of Mont-St.-Michel (r),
white cliffs of Dover (below l),
and French countryside.

the Loire, and white butter and shallots. We stopped for the night at La Rochelle, an old seaport massively fortified with walls and watchtowers against perfidious Albion. At the restaurant we chose, down under an awning by the old port, the staff were desolated, but it was late and they were closed and the chef gone home. If we could manage on some cold lobster and mayonnaise, and simple steaks and fries, and Camembert to follow. We decided we could so manage.

And so it went. In Bordeaux we spent a day at one of the great wine châteaux, much of it consuming a lunch that began with eggs in aspic and Roederer Crystal champagne, and continued through fish, steak with some of the good '55, to some light-as-a-feather confectionery worth about a million calories a munch and served with a glass of ice-cold Château Y. Then coffee and Armagnac, and they drove us, torpid but triumphant, to our hotel.

In Carcassonne we ate cassoulet. Then we flew along the Mediterranean coast, over the Camargue, a marshy plain of reeds and bird sanctuaries that is France's Wild West, where they herd cattle on horseback; over the many mouths of the Rhône; past Marseille; past Toulon, a naval port where an ancestor of mine, a French general besieging the place in a civil war two centuries ago, sent for a young lieutenant of artillery whose gun-laying had been notable and asked his name ("Bonaparte, sir, Napoleon Bonaparte"); past the red rock mountains and clustering red-tiled hilltop villages of Provence. Our Cessna 172 with its high wing and big windows gave a panoramic view of the lovely landscape drifting by below. Our destination was St. Tropez, whose little airport at La Mole lies quite deep in the bottom of a valley surrounded by aromatic piny mountains, and requires a careful approach with a steep turn onto finals. You are advised not to land here in a mistral, the enormous mountain wind that blows here sometimes, and indeed I would not want to try it, but this evening it was calm and hot. A taxi took us to the Byblos, a surreal hotel in town that is a French architect's idea of clustered Saracen battlements (clustered around nothing more warlike than the kidney-shaped pool), where we consumed lamb and beans and *vin rosé* by candlelight under a zillion pin-

prick stars, and tried not to think about the price of our rooms. It's only money, we told ourselves.

The weather was fair all over Europe, but I had to return to New York, and I flew back to London in the 172 in a day, feeling France and I had discharged our duties well. It is a fearsome responsibility, being in charge of someone else's honeymoon. What if it had gone wrong?

The pace of that trip had been easy—an hour's flying in the morning, an hour after lunch—and that is important. Fly too long and it ceases to be a vacation. I once flew ten hours and fifty minutes in a day, from El Dorado, Arkansas, to El Paso, against a stiff wind in a slow plane—a long time to be stiff and hot and weary of the bumps and the engine noise and the the need to be constantly watchful for other traffic. Little pleasure in it, after the first hour or two. That vigilance is vital, always, was brought home to me the very next day, somewhere north of Tombstone, Arizona. After seeing small sign of man in the desert landscape all day, and not a single plane other than my own, I was growing drowsy in the noon sun. When a shadow briefly darkened the map I was studying, I glanced up to see a Comanche go by just a hundred feet above me! I have twice flown coast to coast and back in little Cherokees, and I would not do it again. It's just too far—four days going west against the wind, three days back, flying all day, even with fair winds and weather. An adventure, true, and I am glad to have done it. You meet some marvelous people at little airports in out-of-the-way places, and you gain a unique insight into the geography of the United States. But three thousand miles is too long to enjoy sitting in the same seat in a small plane. Five hours in a commercial jet makes more sense—or ten days in an automobile, *really* seeing it all.

At the other end of the spectrum is what the owner of an old fabric-covered Tripacer once described to me as "the pursuit of the $25 cup of coffee." That was his sort of flying. Untie the old machine, start her up, and fly twenty minutes to a neighboring airport, have a cup of coffee, and fly home again. Total cost, $25. Net gain, one cup of coffee. In between lies the most useful and comfortably horizon-broadening type of lightplane flying:

126

Banking plane tilts 13th-century
Dover Castle slightly askew (l), while
Salisbury plane attracts
more glances than Stonehenge.

the day or maybe weekend trip of one or two hours out, the same back. If you take your family, this is, with kids, about the maximum that is comfortable for you and them. But the scope within this two-hour maximum is wide indeed. From New York City, for example, it gives you Philadelphia, Washington, the Jersey shore, East Hampton (easily the most beautiful old township in the U.S.), Montauk Point (eastern Long Island being agonizing to get to by car in summer), Cape Cod and much of New England, the Finger Lakes and the Catskills, southern Vermont, the Amish settlements of Lancaster County, three quaint islands—Nantucket, Martha's Vineyard, and Block Island—and so on. And from London I can reach Paris, Brussels, Amsterdam, Normandy, and the marvelous little seaside towns along the French and Belgian coasts. Much of my weekend flying is going to air shows, and why not? No traffic

jams, usually no gate fees for visiting pilots, and I am sometimes nearly home before the four-wheeled folk are all out of the car parks. And in Europe, at least, many motor-racing circuits have associated landing strips.

Weather being the bane of small-plane flying, before a shortish trip you can usually study the weather situation on the morning you plan to go and get a fair idea if it will hold up till you plan to return. If it looks at all doubtful, you simply make other plans, boxing the kids' ears if they express disappointment. And there's nothing like wending your way home at 1,500 feet on a fine Labor Day or Thanksgiving evening and watching the bumper-to-bumper crawl on the highways below. There's little enough traffic in the skies, anywhere, and the only speed limits are in terminal areas. And 250 knots, or 287 mph, is no hardship.

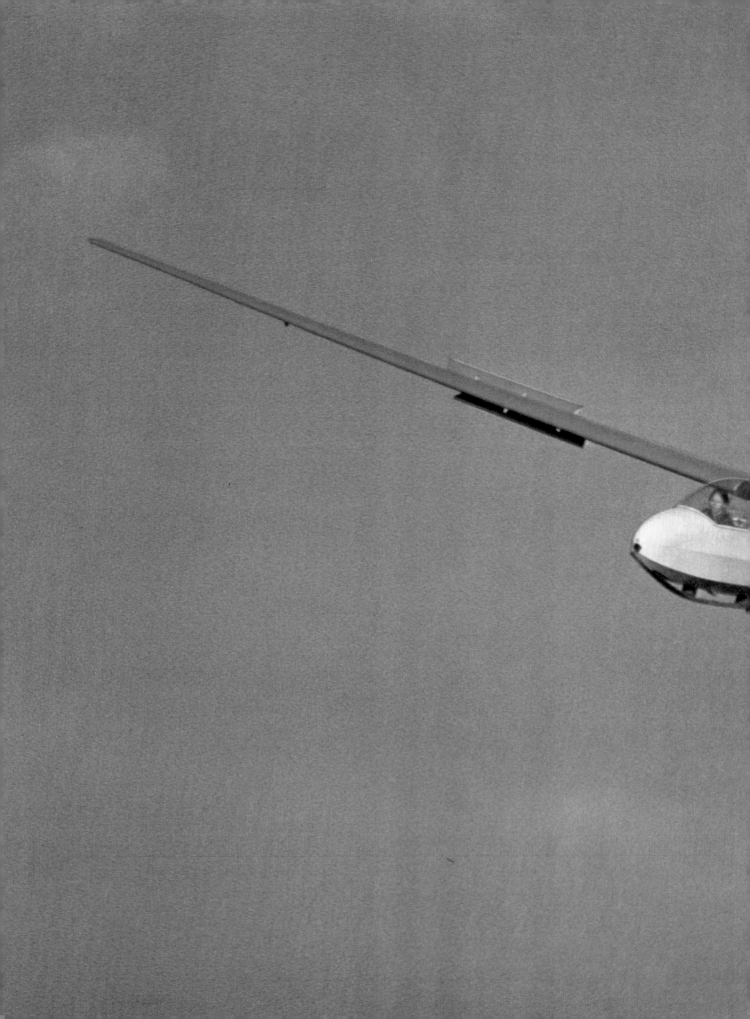

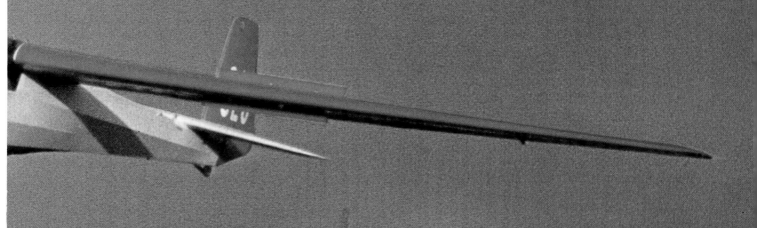

6

Slope-soaring at Fayence. Glider (here
with tow plane still attached) enjoys powerful
updrafts that make this site in southern
France a favorite with soarers. Below: Tow plane
accelerating for takeoff. Light
glider is already airborne.

turn, rudder to hold the nose accurately just below the horizon while in a steep turn, and rudder again to straighten up when rolling level out of a turn. Aerobatics (in a sailplane that is approved for them) are mostly very quick and tight: tight loops, a quick spin, but slow hammerheads, and the most ponderous slow rolls imaginable. The most delicious thing about glider flying is the quiet. At last, an airplane in which you can talk with your instructor at an ordinary conversational level. Finally, a glider burns no precious fossil fuel, generates no smoke, makes no sound above a murmur.

I first soloed in a glider at Wurtsboro, New York, one pale blue spring evening. There'd been little lift all day, but just as I was launched a feeble breeze arrived and with it there formed a tiny wave in the lee of the hogback hill nearby. It was nothing you could ascend in the old Schweizer trainer, but with care you could maintain height. I hovered for forty minutes, till my instructor, who wanted to shut up shop and go home, started flashing the lights of his car to get me down. I thought I'd been cunning enough to stay up forty minutes on a first solo in such weak conditions, but all he had to say was that 600 feet was too low for a beginner to be trying to soar, even right above the field. Well, he was the boss; but I still enjoyed a feeling of quiet, sneaky pride. Soaring, which lets you cheat gravity, the weather, nature itself, is quite definitely the sneakiest kind of flying.

The most bewitching soaring site I know is Fayence, in southern France. It is in the foothills of the Alps, behind Cannes, and ten minutes' flight from the Cannes/Mandelieu airport or an hour's drive from the beaches, up switchback roads through a rocky, pine-tree landscape so cloaked in wild flowers that they seem waterfalls of color at the roadside. The gliding field itself lies up a sandy path through a vineyard and is boarded by fields of lavender, neatly planted like rows of purple pincushions. (The wild flowers fade soon after midsummer and the airfield then becomes a near-desert of brown grass and thistles and flints.) The landscape rises sharply behind the airfield, through terraced olive groves set among palm trees and dark cypress to a steep hilltop crowned by Fayence itself, a snail-

like whorl of ancient terra-cotta roofs that crowd the road as it winds to the summit. So steep is it that the village seems to be tumbling down the hillside. At the top there is a little square with a bubbling fountain in which women launder clothes. Above it is a small hotel with not a bathroom to its name, but a coveted star in the *Guide Michelin* for its cooking of such local delights as saddle of lamb broiled in herbs, chicken with pine nuts, and thrush pâté.

On summer evenings great clouds of swifts scream round the housetops, joined often by the last whispering gliders of the day. When all the thermals of the plain seem to have died, the last weak lift is to be found bubbling up from these still-warm roofs, and the airfield lies handily below the ridge when even that is gone.

Finding lift at Fayence is not a frequent problem. That blazing Mediterranean sun generates vigorous thermals all summer long. Behind the village higher ridges rise steeply toward the high Alps proper, and a daylong sea breeze flowing up them makes for glorious slope-soaring. The ridges are often steep, rocky, fissured by dry ravines, and it is a challenge to follow their contours. In slope-soaring the best lift is closest to the slope; flying these convoluted contours takes concentration.

One misty morning an instructor at Fayence showed me how to slope-soar along one face of a big ridge inland of the field, then round its corner and continue slope-soaring on the other side, a miraculous trick since you would expect only downdrafts in the lee. This was, he explained, a local phenomenon called the *discontinuité*, a strange shear line where the southerly Mediterranean sea breeze encounters the prevailing northwesterlies of central France. The old two-seat Wassmer Bijave we were in was something of a clunker, performance-wise, but so abundant was the lift with the sun heating the rocks that soon the ridge started to sink below us, and after twenty minutes we were a couple of thousand feet above it and being sucked powerfully toward the dank cumulus cloud above. We had so much altitude we were able to head off into the high hills to execute a whispering buzz-job on a mountaintop radar station of the French air force, where my instructor had a friend serving as a technician.

137

Gliding near Shawangunk Mountains, near Wurtsboro, New York. Moment before takeoff (bottom l). Left: Picturesque landing on strip between cornfields. Lens distorts length of wing (below), but not by much. Glider folds neatly onto trailer for convenient ground transport.

But the lift that Fayence is famous for is the wave that builds when the mistral blows. The mistral is a torrent of cold, dry, northerly air that blows only when conditions are right—when pressure is high in the Rhône Valley to the northwest and low in the Gulf of Genoa to the southeast. (The chinook in the Rockies' lee is a similar wind.) The mistral may blow 40 knots, making the trees bend like ballet dancers and setting everyone's shutters banging. It may, according to a local saying, blow for three hours, or three days, or three weeks, till tempers start to fray. But it is fabulous for gliders. When it starts, the haze suddenly clears, leaving a sky washed clean of anything but pale blue. Stationary clouds shaped like flying saucers start to build as the rushing air forms giant waves in the lee of the mountains. Everyone who can gets himself launched into the air in a glider, although the low-level turbulence may be horrendous. In no time at all you are showing three, then five, meters a second *up*, and the lift becomes perfectly and eerily smooth as you climb, pointed into that huge wind and nearly motionless with respect to the landscape below. (And if you fly into such a lift in a two-seater with a French pilot, you will discover that Frenchmen really do say "Oh la la" when they are excited.)

The first day I went to Fayence, as a twenty-year-old kid, the mistral started in the evening. My instructor took me aloft in a trice, a perfectly smooth lift to 12,000 feet (3,658 m). Only the proximity of the high-altitude airway going toward Nice airport, our lack of oxygen equipment, and, frankly, the cold (near freezing, from 75° F [23.9° C] on the airfield; we were wearing summer shorts and T-shirts) stopped us from going higher. The lift was there, all right. The view from our little sailplane was awe-inspiring. We could see the Côte d'Azur from Monte Carlo in the east, sitting below its mile-high rock backdrop, to St. Tropez and beyond in the south. A hundred miles to the north were the white glacial peaks of the high Alps. All this on my first sailplane ride ever!

The mistral blows most vehemently, and the waves are at their strongest, in the winter. It is principally then that the hordes of soaring enthusiasts from flat and waveless northern Europe (particu-

larly the Germans) descend on Fayence to experience the kind of effortless lift they can otherwise only dream about.

But even in summer it can be interesting. In an ordinary powered lightplane in that area I have been throttled back but still climbing at 1,500 feet per minute one moment, then plummeting at 900 despite full power the next. (And nearly thrown on my back by the turbulence in between!)

Although wave lift is noted for its smoothness, there can be extreme turbulence associated with it. The worst is around the rotor, the low-altitude rolling cloud that forms in the mountain's lee, more or less level with the tops of the peaks. Even higher up the turbulence can be fierce; the clue here is when the saucer-shaped clouds that mark the up parts of the wave (lenticular clouds, or "lennies," as soaring pilots call them) begin to assume a ragged, shredded appearance, particularly at their downwind edge. I've seen really ragged lennies only once from the air, going from Las Vegas to Reno one afternoon in a powered Cherokee lightplane, when a huge wind suddenly got up. Even though I went as high as I dared without oxygen—13,000 feet (3,962 m)—I had an awesome and frankly terrifying ride. My route that day took me close to the home of perhaps the world's most famous wave, that in the lee of California's High Sierra mountains. Paul Bikle's absolute world altitude soaring record was set here. He is convinced the wave continued well into the stratosphere and that he was going up at 500 feet per minute when he quit—the reason for quitting being that to go any higher and survive he would have needed a pressure suit, for oxygen alone would no longer be sufficient.

Waves form wherever there are mountains and the wind across the ridges approaches 40 knots. Pilots report fine wave-soaring in the lee of the Southern Alps of New Zealand and great possibilities in the Himalayas, though few have been able to try them. I have one English friend whose lifelong ambition has been to soar above the 30,000-foot peak of Mount Everest. I have another friend, a very senior airline captain, who says the only encounter with extreme turbulence in his long transport-flying career came when he was crossing the Andes at

33,000 feet in a jet airliner, with a 40-knot wind across the peaks and perhaps a jet stream blowing, too. In a clear blue sky the air suddenly went wild. In the next ten minutes he saw, when his eyes could focus on the dancing instrument panel, altitudes of from 35,000 to 29,000 feet (10,668–8,839 m). He is certain that at one point his big airplane went supersonic. At another he remembers the stick-shaker warning that he was close to a stall. The flight recorder readout later showed accelerations of five g; and even this may not have been the true extreme, for the recorder itself broke some ninety seconds into the encounter. Although my friend landed his plane safely, its wings were permanently deformed and the vertical fin was loose at its attachment to the fuselage. Interestingly, other flights at similar altitudes along that airway ten minutes earlier and later met nothing extreme in the way of turbulence. One wonders whether anyone would have believed such a tale, had there not been the mute evidence of the damage to substantiate it.

Who were the first men to soar? The answer must be Wilbur and Orville Wright, in their experiments with simple biplane gliders in the early days of this century—experiments that led to their much-better-known powered flights. With their 1902 glider they made almost a thousand glides over the sandhills of the North Carolina coast, the best covering 623 feet (190 m) and lasting nearly half a minute. Undoubtedly, the wind lifting over the big dune they launched from helped sustain them. Indeed, they noted that there were moments when the glider seemed to be rising. They abandoned gliding for powered flight, but in 1911 they returned to Kitty Hawk with a new glider, similar to their earlier machine but with a conventional rear tail. With it they made one soaring flight of nine minutes, forty-five seconds in a strong wind. This was a world record that stood for a decade. In that same year university students in Darmstadt, Germany, began making hang-gliders, initiating a tradition that continues in German universities to this day. Hang-gliding was not new. There had been several experimenters in the 1890s; two of the best-known, Percy Pilcher of Scotland and Otto Lilienthal of Germany, had been killed in hang-gliding accidents. The Darmstadt stu-

Left: Teardrop fuselage and knife-edge wings are characteristic of contemporary high-performance fiberglass gliders. Below: Lift is strongest in heat of day. Toward evening, as sun starts sinking, air does, too. This does not discourage pilot being towed across sunset sky in search of rising air for one last soaring flight.

Cessna L-19 Bird Dog (r) is an ex-military
aircraft that combines high power
with slow speed, an ideal blend for glider
tow plane. Far r: Patient pilot, like
skier lining up for another kind of tow,
waits his turn to fly. Bottom: Wassmer
Bijave is an elderly but capable
two-seat French trainer.

7

Racing

Reno at race time: northern Nevada is suddenly host to twenty thousand fans converging on the desert town in a mass migration. Mostly they are from California, but many have come, by airlines and private planes, from all over the U S of A, and a tiny handful, all jet lag and excitement, have flown in from Europe.

In air racing Reno is the Big One (although Mojave, in Southern California, is a strong rival), and it always happens in September, a confluence of several seasons in the desert. It may still be hot; one year there were undesertlike rains and thunder; once I saw snow sprinkling the mountaintops, shining yellow at dawn and blue at midday with a clarity that belied their distance from the town.

Although it is the casinos that put up the big prizes, Reno itself has never seemed to me to quite understand a fascination for speed in the air that surpasses enthusiasm for more sinful pastimes. Though the town's motels and restaurants are packed with fans, the neon glow of the gambling halls attracts an entirely different crowd, and the re-marriage chapels must surely count air racing for worse, rather than for better. Only the Oldest Profession seeks to embrace the Newest Sport; I have seen the "girls" from an outlying desert "ranch" that has its own landing strip mingling with the crowds, handing out printed fliers to the fliers.

The races are held over three long days at an old, half-abandoned Air Force base north of town, on a flat and nearly level plateau named the Lemmon Valley that seems to be ringed by distant mountains, making it a fine vast amphitheater. On the program there are aerobatic demonstrations as well as class races for sport biplanes, tiny 100-hp monoplanes, and big old 600-hp World War II AT-6 training planes. But it is the Unlimiteds that crown the day. Theirs is the fastest sport on earth, for they race at up to 450 mph—or 660 feet a second!

The class is not quite perfectly named, for jets and prop-jets are excluded. An Unlimited is anything with a piston engine—which means old World War II fighter planes. (A couple of World War II twin-engined bombers, and even a four-fan DC-6 airliner, have competed, but this, I feel sure, was more for the amazement of the fans than with any real expectation of winning.)

As the planes sit in a long line before the race, you will notice that 51s predominate—North American P-51 Mustangs of the type that once escorted the Eighth Air Force's bomber fleets on their daylight raids over the Third Reich half a lifetime ago. There is also a handful of Grumman F8F Bearcats, Navy fighters too late for combat in World War II, but which flew (in French and Thai hands) over Vietnam. You may also see an occasional Lockheed P-38 Lightning, P-63 King Cobra, Republic Thunderbolt, Vought Corsair, or British Hawker Sea Fury. Mostly they are stock, hardly altered airplanes painted in their old military colors. These aircraft do not win. It is "specials" that are the hot ships: Bearcats and Mustangs with cut-down wings, tiny teardrop cockpit canopies, and reworked engines of enormously increased power—and unreliability, as the actual racing often reveals.

Parked in the pits, they usually seem to be floating in a dense sea of fans. The top halves of their propellers rise above the mass like brandished scimitars. The spectators clear away before the start. The racers depart one by one, propellers slowly churning till the engines fire up in clouds of blue smoke. They taxi out, run up, and take off in a thunderous roar, then circle lazily while settling into a big echelon formation alongside the pace plane, a stock 51 flown usually by Bob Hoover, an ex-test pilot and air-show regular. If you have a little VHF radio you can tune in and hear the race pilots checking in with Hoover, and Bob calling them into proper alignment in his Tennessee drawl: "Pick it up a little, Race 16, you're dragging. Okay, that's good. Hang right in there where you are, everybody. . . . You're all looking fine, mighty fine. . . . Now don't lose it, anybody, we're going in."

In the distance you see Hoover leading them out of their circle against the distant mountains and heading in toward the start line, accelerating and losing height. They buck slightly in the turbulence, but still hold their echelon positioning against Hoover's yellow Mustang. Near the start he calls, "Full throttle, gentlemen, full throttle!" and a moment later he banks hard left and pulls up inside the race course, calling, "Gentlemen, you have a race!"

Almost as soon as there were airplanes, there was air racing. From Reims in 1909 to Cleveland (preceding pages) or Reno in the 1970s, the quest for speed has gripped all manner of fliers—and spectators (below) watching prerace aerobatics.

He remains airborne for the duration of the race, circling high above the course, which is eight or ten miles long, defined by eight or nine brightly colored pylons where crews of judges check the racers around. (A pylon "cut" costs the pilot a whole lap.)

The whole wide valley echoes with the howl of their engines, which rises to a deafening blare once a lap as they come past the pits and the spectator areas near the end of a long straightaway. Winning average speeds are now around 430 mph, with 480 probably achieved at the end of the straights. The first sign that someone has blown his engine is usually when he pulls up in a zoom, trading speed for altitude while he plans a forced landing on the airfield in the center of the race track. Here Bob Hoover, circling high above, can often help the pilot in trouble, whose forward vision may be obscured by oil covering his windshield. When Mustang pilot Clay Lacy had a runaway prop in a race, Hoover was able to point out, over the radio, a runway he could land on. And then when Lacy's 51 wouldn't slow down because of the prop, he could tell him: "There's a fire road that angles off from the far end of the runway—take it." Lacy's 51 went a mile down the narrow track in a huge cloud of dust, but both plane and pilot escaped unscathed. During an emergency the other planes in the race must rise from their usual racing altitude (just above the sagebrush) to at least 300 feet and maintain their order till the stricken machine is down—a holding pattern like that established by the yellow flag at Indy. Most racers get down safely, but there have been tragedies; a blown engine may be followed by a fuel fire.

Unlimited pilots are among the most colorful characters in aviation. Hoover is a former North American test pilot whose air-show routine in the 51 is as spectacular as anything you can see done in an airplane. (He flies similar shows in Rockwell Shrike Commanders and Sabreliners, F-86 and F-100 jet fighters, and once, unplanned and unscheduled, and to the fury and amazement of the Russians, in a borrowed Yak aerobatic contest airplane at Moscow in 1966.) Hoover has bailed out five times in his long career, and once demolished a P-51 right in front of the crowd when he lost power during a low-speed roll at low altitude, flying with gear and flaps down.

Racing at Reno: Overview, with heavyweight Unlimited
(piston-engine) Class, which means World War II
fighter planes, in foreground. Right: Two premier Unlimiteds
are *Red Baron,* a P-51 Mustang with double
contrarotating props and 3,400-hp Rolls Griffon
engine, and Darryl Greenamyer's *Conquest 1,*
a much-modified Grumman Bearcat. These planes race at
some 450 mph (724 km) over 8–10-mile courses.

Open-cockpit, fixed-gear, fixed-pitch prop
sport biplanes produce 200-mph speeds on modest
budgets. Wild paint schemes and
unorthodox design layouts are notable in
this individualistic class, an offshoot of the
fly-for-fun, home-built plane movement.
Below: At start, in full flight,
and (opposite) a gung-ho race in progress.

Darryl Greenamyer, meanwhile, has been working on a project altogether wilder and faster. He has been building up his own F-104 Starfighter with parts collected from various places. The 104 has always been his favorite, he says, and he considers it better than most more recent military jets. With his 104 he plans an assault on some world turbojet records, those for altitude in level flight and time-to-altitude. He has always maintained that the pursuit of world records tends to be political. Congressional appropriations must be justified, so any new fighter plane must be made to seem faster than its predecessors. The 104, he claims, has always been underrated, and is still capable of breaking several world records. Imagine: your own private Starfighter! It is surely the ultimate hot-rod project.

The T-6 Class in air racing is confined to one aircraft type: supposedly stock North American AT-6 training planes (called SNJ by the Navy and Harvard by the Canadians and British). It is the noisiest racing you could hope for. The propeller tips of these birds go supersonic at full power, producing a shattering drone that rattles everything loose for miles. The engines they all have are R985s, Pratt and Whitney radials of some 600 hp. Old-fashioned radials and proud of it. One T-6 owner has lettered on the cowl of his plane: "If God had intended aircraft engines to have horizontally opposed cylinders, Pratt and Whitney would have made them that way." Hard to argue with that. I can't help feeling that these engines are getting less and less "stock" as the racing years go by. Winning speeds, some 180 mph a few years ago, are now over 210. You don't gain 30 mph just by thinking about it.

In the same way that Formula 2 and Formula 3 motor racing are breeding grounds for kids trying to make it to the big time, T-6 air racing has been where some famous Unlimited pilots started. There is another parallel with motor racing: while the Unlimited pilots have their hands full staying out of each other's way and waiting for their engines to blow up, and in consequence fly pretty conservatively during the actual races, the T-6 boys are hard chargers, and get closer to the ground, the pylons, and each other than you would believe. Even the occasional tragedy (they happen) doesn't seem to cramp their racing style. While their speeds are half those of the Unlimiteds, you'll see the most spectacular flying from the T-6 jockeys. Not so much toward the finish of a race, when they are getting spread out around the course, but in the early laps, when they come round the pylons all bunched. Sometimes I can hardly stand to watch.

Formula 1 (oddly named, since there aren't any Formulas 2 or 3) was introduced immediately after World War II, and was intended to be an inexpensive way to go racing—as well as a reaction to the overpowered slaughter that prewar air racing had begun to be. The requirements defining the class include: a minimum 66 square feet of wing area; at least two wheels of a minimum size, fixed gear, and brakes; fixed-pitch propellers; 500-pound minimum empty weight; and a "stock" 0-200 Continental four-cylinder engine nominally of 100 hp. A fair amount of engine "improvement" is permitted, and the rules defining this become more complex every year. You can do quite a lot of balancing and matching and polishing; for instance, you can machine and polish three pistons to match whichever is the lightest in the original set. Prop-shaft extensions are permitted, as are modified sumps and oil tanks. You must use the stock carburetor, but you can improve its air supply and engine cooling arrangements as much as you like. In consequence F 1 engines are very highly tuned, indeed, typically turning at up to 4,000 rpm against the stock 2,750, and producing around 150 hp—half again as much as the standard. F 1 rules have been modified again and again down the years and are full of waivers and exemptions for the benefit of airplanes that have been racing a long time. And while you need a fair amount of time, a good machine shop, and quite a few thousand dollars to produce a competitive F 1 racer, the rules have at least prevented the class from turning into the kind of triumph-of-pure-money that the fastest Unlimiteds undoubtedly represent.

There are more F 1 races than any other kind. They have spread to England and France, as well as to the eastern and central parts of the United States. They race around a standard six-pylon course of two and a half to three miles (the same as that for the

Formula 1 was conceived as inexpensive racing for small but lively planes, now offers more races than any other class. Counterclockwise: Bob Hoover, ex-test pilot and aerobatic expert, performs his famed air-show routine in Rockwell Shrike; crew member with cockpit canopy for *Rivets;* psychedelic skin color; rolling all-plywood *Stinger* to start.

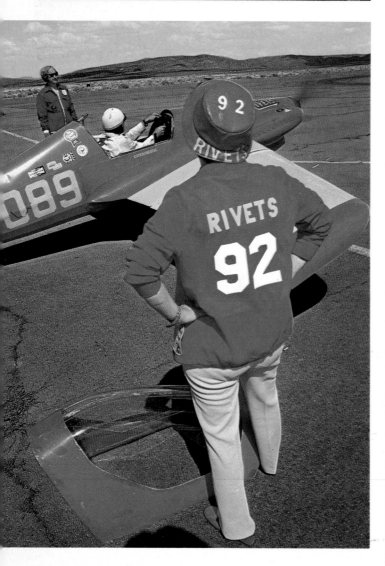

Formula 1 heat—usually six planes—
strains at start, awaiting flag signal. Engines
are wound to full power, pit crews
hang on tail to help brakes hold plane in place.
Below: Judy Wagner, one of few women to
make championship field at Reno and
Mojave, in her *Wagner Solution*
Formula 1 racer. Plane's long,
tapered wing design makes it particularly
successful at taking turns, a critical
feature in Formula 1 racing.

AT-6 and sport-biplane classes), which keeps them always in view of spectators. Race speeds are up to 240 mph, and the impression of speed is heightened by the tiny size of the aircraft. Though unconventional layouts such as rear-engined pusher configurations are allowed, most F 1 racers are conventional-looking airplanes with tractor propellers, tapered mid-wings, and rear tails. The winningest racers in the class are usually the oldest, aircraft that have been progressively modified and improved, achieving contemporary high speeds only after unstinting effort.

To an amateur student of aerodynamics it is

fascinating to note the little differences in design detail among the fastest racers, all working toward the goal of minimum drag at up to 270 mph and four g. A great deal of professional scientific design knowledge is, happily, *not* required to do well in the class. The two most successful designs are the work of a manager of a business flying operation and a guy who, until he retired, ran a little airport in upstate New York. There's a lot of spit and polish in it, too. In an effort to remove the last unnecessary ounce of parasitic drag, F 1 airframes are smoothed out with balsa and body putty and fiberglass and then painted and polished till they gleam.

164

Racing aside, F 1 airplanes are delightful to fly, with rates of climb of up to 3,000 feet per minute, cruise speeds of 200 mph, and tantalizingly light, powerful controls. (They often have almost full-span ailerons, giving them an exhilarating rate of roll.) But you had best not be claustrophobic if you are going to fly one, for their cockpits are minimum size. Be sure to wear earplugs. The engine noise inside that bubble canopy is shattering.

An F 1 heat or race is typically among six airplanes. On a flag signal they make a race-horse standing start from a line across the runway. Since runways are often not very wide, the commonest ar-rangement is with four airplanes on the actual runway and two lined up on a parallel taxiway. As the time approaches for the start, pilots wind their engines up to full power, with pit crews hanging on to the tails to help the wheel brakes hold the planes back. The lightest planes get off and around the first pylon first. A good start is of less importance here than in automobile racing, for overtaking is easier; indeed, one F 1 pilot who has won an extraordinary number of races invariably gets off to a slow start and hangs back, flying high and wide, often until the last lap, when he dives down and tries to take the lead. But whoever is leading does have the eas-

Tiny F 1 racers hit 240 mph (386 km) in
3-mile sprints around standard 6-pylon courses.
Only Unlimiteds fly faster. No. 99 (l)
is Cassutt RG-5 *Scarab* (No. 591A, whose tail is
being held on p. 165). No. 92 (below) is
Rivets, designed and built by its highly
successful pilot, Bill Falck (bottom).

167

Size of little F 1 racer can be inferred
from men studiously ignoring it. Plane they are
watching is F-4 Phantom trainer-bomber,
part of USAF's Thunderbird team.

iest time of it, for he flies in undisturbed air and can pick his ideal line around the three pylons at each end of the course without having to worry about avoiding other racers. Further back in the pack the turbulence can be fierce. It is caused not by propeller slipstream as you might suppose, but by vortices from the aircraft wing tips. High-pressure air under the wings tends to spill round into the low pressure above them, streaming back from the tips in tiny twisting tornadoes. These vortices are fiercest when the pilots are pulling g around the pylons, and they manifest themselves to following aircraft as a sharp rolling moment. It is best to fly higher and outside the airplane in front; if you get into the vortex from his top wing as he banks to turn, it will merely try to roll you out into level flight. But the vortex from his lower wing will try to roll you on your back—at maybe fifty feet of altitude. *That's* scary. The vortex seems to tug and tweak your airplane, and the stick may twitch in your hand, almost as though some living thing were in the air.

Formula 1 may be the lowest-powered racing class, but it inspires airplanes whose speeds are second only to the Unlimiteds.

Some of the freakiest-looking flying machines in air racing are to be seen in the sport-biplane class, an offshoot of the fly-for-fun home-built plane movement, which spawned a host of open-cockpit biplanes generally powered by versions of the Lycoming 0-290. This is a World War II device built like an aero engine, but intended for ground power units. Later it became available at military surplus prices far below those of new aero engines, though still capable of being made, with a minimum of modification, into a perfectly serviceable aircraft engine. Like F 1, sport biplanes are something of an Anyman's air-racing class, offering fine sport at speeds of over 200 mph and on a reasonable budget. This, reader, is a performance triumph, when you remember that the 0-290 was originally a 130-hp unit, and that a biplane by definition has all kinds of inherent drag from aerodynamic interference between the double wings and struts.

There have been drag races, too: from a standing start, once around the course, and over the finish line. And jet racing; jets might seem the obvious

ultimate in competitive speed, but they are incredibly expensive and mostly in the hands of the military, who frown on the frivolity of simple low-altitude racing. The military did enter a few jet races just after World War II, but not many have been flown since. The tiny handful of ex-service airplanes in civilian (*rich* civilian) hands are too few to make good sport, and their speeds vary too widely to make much of a race. At Mojave in 1974, for example, Bob Hoover in his F-86 Sabrejet was lapping at 525 mph, 160 mph faster than the de Havilland Vampire that was the slowest of the six competitors. The competitors had to fly a slowest-off-first handicap race that still contrived to be a poor spectacle for the fans. Another problem is that g forces around the turning points tend to approach the limits sustainable by human pilots at 500-mph speeds. Jet racing, perhaps on a class basis, where engine power is severely limited, may indeed have a future, but cost remains an insuperable barrier.

A much-touted class that may yet produce excellent sporting within a reasonable budget is Formula V, restricted to racers powered with Volkswagen auto engines. A few have flown, but VW racing in the air is still a future thing.

Air racing is almost as old as the airplane. The very first aerial race was held at Reims, France, in 1909, just six years after the Wright brothers' first flight—and it was won by a visiting American, Glenn Curtiss, at a tiny 47.65 mph. The first race in the United States was at Compton, California, just five months later. I have never been able to find out who won it, but I can tell you that M. Louis Paulhan of France managed, at that meet, in a kind of reverse transatlantic challenge, to raise the world altitude record to a breathtaking 4,165 feet (c. 1,270 m). Flying for the public was a fearsomely dangerous business in those days. Thirty-seven professional pilots were killed while racing or stunting in 1910. Perhaps the greatest days of air racing as a public spectacle were in the twenties and thirties, in the contests for the Bendix and Thompson trophies and the Schneider Cup. The fact that so many men died in these events undoubtedly helped draw the crowds. The carnage is over in air racing, but the spectacle remains.

8

Home-built Airplanes

To sport-flying enthusiasts, many "store-bought" airplanes seem somewhat bland in their handling. Not, of course, the big twins: the Aerostars and Beech Dukes and Navajos have superb handling qualities and tremendous performance. They are also tax-loss barouches, way beyond the means of ordinary aviators. (There was a nineteenth-century English lady named Mrs. Beeton who wrote a cookbook containing a recipe for jugged hare that began: "First catch your hare." For one of those big twins, first make your million.) It's the cheaper single-engined types that some purists affect to despise. Consider the world's best-selling lightplane, the Cessna 172. Something like twenty-five thousand have been built and sold, and it's easy to see why: the 172 is an economical, 150-hp family four-seater, a stable Mabel that is easy to fly and easy to land, with good baggage room, comfortable seating, and respectable sound-proofing. But a superresponsive hot rod it ain't. One reason it isn't is that "family" airplanes of this sort all have ram's-horn-shaped yoke controls, which project out of the instrument panel, leaving an unobstructed floor space so that women wearing skirts can get in and out of the airplane comfortably. A stick control would give you better leverage, shorter cable runs, and fewer pulleys in the system. Still, a yoke is undoubtedly the right merchandising choice.

Cessna did put a very powerful elevator control on their early Cardinal, making it, I thought, a really delightful machine to fly. But a small percentage of pilots couldn't handle this, and would get the airplane porpoising on landing, even to the point of wiping out the nosewheel. So the Cardinal's elevator was substantially modified. To some extent the commercial airplane designer has to design for the lowest common denominator of pilot skills. The United States' stringent product-liability laws are a further encouragement to the designer to try to make airplanes practically idiot-proof.

There is, surprisingly, less interest in gadgetry for its own sake in small airplanes than there is in many mass-produced automobiles. Except that there is hardly a lightplane design that does not have plenty of space in its instrument panel for the installation of expensive avionics gear. Avionics, in fact, sometimes accounts for a third of the airplane's price. Many family four-seaters leave the dealer's hangar with enough equipment for a full instrument flight—which is often far more than the buyer can handle. I sometimes wonder if people really want all those radios. Or are they, perhaps, afraid to say no to the salesman, or to seem cheap? As with automobiles, no one ever seems to settle for the basic model or the cheapest option list.

As far as merchandising considerations go, there is no doubt that the lightplane manufacturers know their market well. Yet I wonder if, in manufacturing mainly easy-to-fly family four-seaters of quite modest performance, they are in danger of forgetting the romance, the excitement, the traditions, history, myth, and simple magic of flying. They know better than to build airplanes for dreamers; but was it not dreams that got us all flying in the first place? Occasionally a manufacturer shows glimmerings of insight. One, for example, painted several of his airplanes like Royal Air Force Spitfires and Curtiss P-40 Flying Tigers for a photograph, and then decided to offer those designs as options you could buy. But underneath the gorgeous colors, the airplanes were still family sedans, with plenty of room behind the panel for expensive avionics.

The manufacturers, then, take care of the mass market. Those of us who do not conform to their marketing profiles, and who do not want aerial Buicks, simply have to roll our own. Hence the home-built airplane movement, also called "sport flying" in recognition of its one unifying factor: its attraction for people who insist that flying is not just a means of transportation, but real sport. The innate conservatism of the big manufacturers has been nicely described by Molt Taylor, a maverick engineer who is a force in the home-built world, having pioneered the flying automobile (its wings and tail detach for highway operation) and amphibious airplanes for home construction. "The big manufacturers couldn't care less whether they are building teats or tiddlywinks," he asserts. "The motivating force of any corporate enterprise is profit. . . . They don't care whether an airplane has ten wings or two. They're interested in making a buck."

Taylor also sees the manufacturers as being in unholy, if unintended, alliance with the finance

Preceding pages: Admirable flight characteristics of 1929 Brunner-Winkle Bird make it sought after by restorers. Below: EAA Convention at Oshkosh. While spectators concentrate on overhead attraction, pilots prepare for demonstration of formation aerobatics in BD-5J, jet-powered version of BD-5, popular home-built designed by Jim Bede.

houses. "When you buy a production airplane the first thing you've got is a finance problem," he says. "You can't buy any kind of airplane today that doesn't cost you two hundred fifty dollars a month in financing. Add to that another forty to fifty dollars in hangar rent, and more in annual inspection, radio repairs, engine maintenance, fifty cents a gallon fuel, and the damn thing will fifty- and seventy-five-dollar you to death." Whereas with a home-built, you may have to shell out a couple of hundred dollars for the plans initially, but thereafter you can buy materials, parts, the engine, and so on, only as and when you wish. There is no continuing financial commitment. "There is also," says Taylor, "the convenience of being able to do your own maintenance, service, and inspection."

Taylor engineers his own designs for home-builders to meet many of the FAA certification requirements for production airplanes. Otherwise he

is, in the words of *Air Progress* magazine, a person who "fiercely denounces the whole federal regulation process as a grievously inhibiting factor in the growth of aviation." "I don't think that certification adds one dime's worth of value to the airplane," he grumbles. "In a free marketplace the problem would be taken care of without government regulation, because if you started selling airplanes and the wing fell off, you wouldn't be in business very long."

Home-builts do all have to go through a form of certification, but one far simpler than that required for production types. In most countries the government delegates approval authority to officials of home-builders' organizations, such as the Experimental Aircraft Association in the United States. These inspectors examine the craft at various stages during its construction. If the design is a new or unfamiliar one, they appraise it at the design stage as well. In general, the engineering standards of most

home-built designs are good. Should some inherent defect be discovered in a type, the EAA promptly advises members building that type of the correction required. Yet the quality of the plans offered for sale varies widely, and there are a few poor, ill-conceived, and badly engineered designs for home-builts on the market. In the words of Harold Best-Devereux, who runs the English office of the EAA, "Let *caveat emptor* [Let the buyer beware] be on the lips of the plans purchaser. The world over has glib, self-confessed geniuses selling rubbish by the bucketful. I am always amazed how these con-men and plagiarists get away with it. One I recall could not tell the proud constructor where the center of gravity of the airplane should be, yet he claimed to have 'redesigned' the aircraft, whatever that might mean." The newcomer's best defense against this trap is to seek advice from his local EAA chapter to help him choose a design to match his skills and experience, not just as a builder, but as a pilot, for there are some home-built types that are decidedly tricky to fly for the low-time pilot.

If home-built airplanes are a low-cost way to aircraft ownership, they all make prodigious demands on the builder's time. Even the simplest take hundreds of hours to complete. While it might be possible, if you did nothing else, to fly a few months after starting work, two years is the usual minimum and four or five years is more like it for ordinary folk who work for a living and have other demands

on their time. In consequence, there's a high drop-out rate and a thriving market for the sale of partially completed home-builts. Roughly seventy-five percent are never finished by the person who started them, and some sixty percent are never completed at all. There are two or three thousand home-built airplanes flying, and a purported eight or ten thousand in some stage of construction. The slowest completion time I know involved a Knight Twister biplane, admittedly a complex and difficult design, that took twenty-nine years. Its owner was a USAF officer who, like all military men, kept getting posted and reassigned, so that the Twister had to wait and keep waiting till he could get back to it.

What sustains the home-builder and fortifies his perseverance is the thought of one day flying his finished bird to the mecca of sport flying, the annual EAA convention at Oshkosh, Wisconsin. (Or lesser meets at Sywell, England, or in France, or, for the antique-airplane buff, the fly-in at Ottumwa, Iowa.) Oshkosh in full swarm is perhaps the most astonishing sight in aviation. For a week this ordinarily modest regional airport becomes the busiest flying field on earth, with more than a thousand home-built and antique airplanes flying in from every state, and thousands more visitors arriving in store-bought types to see the fun. The home-builts cover the airfield in neat rows, bright as butterflies, each one tied to stakes at wing tips and tail against inclement weather. Like clusters with like. Here is a row of aerobatic Pitts Specials stretching almost to the horizon; here a row of "war birds"—lavishly restored military aircraft of World War II; here the Formula 1 racers, tiny monoplanes sitting so low on the ground they hardly seem to rise above your knees; here a thicket of autogyros and helicopters. Spectators perambulate among them, with an occasional throng of fans around the newest and most fascinating type—a throng so thick you can hardly see what it is they're interested in.

The answer to why no ordinary airplane manufacturer can satisfy the sport flier's requirements is quickly apparent. It is because their dreams are so different, so varied. Some must have amphibious craft, with boat hulls for water landings and retractable wheels for land-based flying. Some must have rotorcraft, but at a price no manufacturer could come close to. Some like tiny, overpowered biplanes whose handling characteristics can be so squirrely no manufacturer would dare offer them to the public. Not all the Formula 1 racers race; some are built by guys who simply crave an airplane that will go 200 mph on just 100 hp, climb at two or three thousand feet a minute but only burn some five gallons of gas an hour. Many—perhaps most—home-builts are single-seaters. Manufacturers will tell you that the market for anything less than a four-seater (other than the purely training two-seat types) is too small to interest them. Many home-builts have open cockpits, their pilots enjoying helmets, goggles, and the wind on their cheeks in the same way that motorcyclists do.

Some home-builders demand performance that ordinary manufacturers cannot offer, and are prepared to accept the necessary limitations that such performance carries. (One home-built, a tandem two-seater with a 285-hp engine and full instrument capability, nudges 300 mph at altitude.) Some crave the kind of enormous range that Lindbergh had built into the *Spirit of St. Louis*. (Home-builts have flown around the world.) One such airplane is Peter Garrison's Melmoth, a two-seater with 3,000 nautical miles of range, equipped for IFR and stocked with gadgets to aid the pilot. "I am drawn to the thought of flying long distances," he explained in *Flying* magazine. "It is my mountain

climbing, my archery, my test of strength. . . . I like to look at a globe, mentally measuring the great-circle routes and translating them into hours and pounds of fuel. The more remote the route, the greater its magnetism for me. My reason sees such long flights as long sits . . . but in my heart they appear as the door handles of eternity, glimpses of the inaccessible and the sacred." And when Melmoth arrived at Shannon, Ireland, from Gander, Newfoundland, enough fuel remained to go on to Rome.

Nostalgia for aviation's romantic past is the dream that moves many sport fliers, including me. I was born in the 1930s near London's Croydon airport, then still served by wood and wire and fabric biplane airliners. The Battle of Britain filled the sky overhead for three months of my childhood. I was raised on books that told of aerial warfare above the trenches of World War I, of barnstorming in the 1920s and ocean-spanning and earth-girdling in the 1930s. And I learned to fly in an open-cockpit Tiger Moth biplane. As a teen-age student pilot I saw Cantacuzene fly his Jungmeister and was enthralled for a lifetime by his sort of flying. Every antiquer dreams of finding the rusty skeleton of some ancient bird moldering in a farmer's barn, and spending a decade restoring it to its original glory. Antiquers regale each other with mythic stories of finding a World War I Le Rhône rotary in a shed, in a perfect state, still packed in the original manufacturer's grease. (It has happened.) Some build from scratch replicas of romantic designs that take their fancy. Although Piper Cubs and Jungmeisters are no longer manufactured, plans for them are available, lovingly copied from existing airframes. The Hawker Siddeley company in England will still sell you plans for the Sopwith Scouts built by an ancestor business of sixty years ago, and the English magazine *Flight* drew plans of German airplanes captured during World War I and has them still. People even build replicas of airplanes dating earlier than 1914: bird-cagelike Curtiss pushers and Santos-Dumont Demoiselles, Blériot and Antoinette monoplanes, even the original Wright Flyer. I would say the number of pre-1945 aircraft in flying condition is steadily increasing, year by year.

Personally, I have never handled an original

177

Echoes of World War I.
Fokker Eindecker (r) and Stampes
(bottom) modified to resemble
SE 5a's were built for film *Aces High,*
in which author was a stunt flier.
(Eindecker, however, had
modern engine and ailerons.)
Opposite: Modern replica of Caudron G-3,
a French recon and bombing plane;
also the first aircraft to
take aerial photos of Peking (1913).

airplane that was more than fifty years old, but I've passed some grand days in World War I replicas built for moviemaking in such epics as *The Red Baron* and *Aces High*. These craft have modern engines, saving you from the tender mercies of unreliable and hard-to-handle rotaries, but otherwise they generally duplicate most faithfully the peculiar and demanding flying qualities of the originals. With fixed tail skids and no brakes they are inclined—like a puppy with fleas—to try to bite their tails toward the end of a landing run, or even on takeoff. Flexible structures and tiny tail surfaces often make them unstable in flight, with a completely "dead" feel to the controls that takes much getting used to. The machine guns either flash electronically or fire in bursts of flame from an acetylene bottle concealed in the fuselage. And since the only way for a "crate" to crash, as far as a movie director is concerned, is to spin down in smoke and flames, we fly with pyrotechnics strapped to our cowlings and landing gear, to be electrically ignited from the cockpit at the criti-

cal moment. Much of the resultant smoke often pours through the cockpit, so that we land after such a mission looking like Laurence Olivier playing Othello in blackface.

We fly alongside a cameraman hanging half out of a Jet Ranger helicopter, or filming through a periscope projecting through the floor of a Piper Aztec twin, or with a fixed movie camera mounted on our own aircraft, looking straight forward over the nose in wide angle at the airplane we are pursuing, or back over our tail at the machine chasing us. Generally it is this fixed camera work that produces the most dramatic footage. We have discovered that it is possible to fly directly in the wake of another aircraft, as close as five or six feet behind it. This is not something you would ever want to do in real combat, but it looks good, if unlikely, on film. The front airplane has to fly throttled well back, and its wing-tip vortices still tend to roll the pursuing plane out of the way, but with practice you can sit there in reasonably balanced flight. We wear leather coats

179

Tipsy Nipper (top), designed by a Belgian named Tips, has VW engine, is low-cost, easy to build and fly. Burt Rutan's Vari-Eze (middle), canard tandem two-seater, weighs a mere 390 pounds, has many ingenious features built in to assure stability and efficiency. Flying Flea (bottom), creation of a self-taught flier, was briefly popular in 1935. However, design was inherently defective, and few Fleas ever flew; there were fatal crashes among those that did.

and helmets. (In one film I worked on it was brown leather for the Allied pilots, black for the Hun, a nice variation on the traditional Hollywood good-cowboy-on-a-white-horse convention.) And scarves; every movie director is convinced that fighter pilots have always worn scarves trailing in the slipstream, half strangling them. Well, it's his money, so we try.

Modern engines aside, the rest of the structure is usually much as the original, simply because even today this is hard to improve on. The wings have wooden spars, built-up wooden ribs, often with piano-wire bracing, and a doped fabric covering. Fuselages were often constructed of wooden longer-ons and spacers with a deal of wire cross-bracing to make the structure rigid. And while welded steel-tube fuselage frames are preferred today, this technique is ancient, too. Anthony Fokker used it for his World War I aircraft. Welding steel-tube structures can be a problem for an amateur builder. You have to know how to weld, and in some countries, such as Britain, you have to be a government-approved welder before you can get a license for a welded airplane. For this reason some designs offer the amateur builder a structure that is virtually all wood and requires techniques scarcely more advanced than those used to make a balsa-and-glue model airplane. Metal airframes usually employ techniques similar to those used in manufactured airplanes, with rivets the favorite fastener, though metal-to-metal bonding is gaining popularity. Fiberglass, plastics, and expanded solid foams are little used, which seems to me surprising in view of their popularity in boat and commercial sailplane construction. Many designs are available as kits of parts assembled in varying degrees of completeness. The FAA's rule that a minimum fifty-one percent of the work must have been amateur-done for the aircraft to qualify for approval in the amateur-built category allows for this. Most builders purchase some parts, such as landing gear, engine- and wing-attachment fittings, canopies, wheels, brakes, and instruments, from professional shops. Domestic garages are the most popular building place. Cellars are hard to get a structure out of. But I know of one airplane that was built in a second-floor bedroom and lowered carefully out a window in sections for final assembly elsewhere.

A variety of ordinary commercial aircraft engines are used, though they are expensive. Converted Volkswagen auto engines of about 60 hp are popular for single-seaters, though most auto engines, while far less costly than commercial aircraft engines, are unsuitable, being too heavy and turning too fast for propeller efficiency. One popular engine is a 125-hp unit originally built by an aero-engine manufacturer in World War II as a ground power unit and available as military surplus. Because of its ancestry it converts easily to aircraft use.

At their best, home-built aircraft designs can show inventive ingenuity scarcely seen anywhere else in aviation. For example, consider Burt Rutan's strange design named VariEze: a canard (rear-wing, tail-first) tandem two-seater that achieves a maximum cruise of 185 mph on the power of a 62-hp VW engine, and a gas mileage of 48 miles per gallon. For comparison, a production Cessna 150 two-seater is 60 miles per hour slower, despite having 38 more horsepower. VariEze is built entirely of epoxy and glass fiber and plastic foams, and weighs only 390 pounds, against the 150's thousand. VariEze's control in both roll and pitch is provided by moving surfaces on the foreplane, leaving the main wing smooth-surfaced and free of drag-inducing control gaps. Instead of the time-honored central stick there is a little side controller for the pilot, a feature also of Jim Bede's BD-5 home-built design and also the U.S. Air Force's new F-16 jet fighter. Tiny winglets are mounted at the wing-tips below the main rudders to give lateral stability and reduce the drag from tip vortices. While the main landing gear legs are fixed, the nose leg retracts to reduce drag in the air and on the ground to allow the occupants to dismount more easily, and to prevent the airplane from sitting back on its tail fins as it would tend to do without the pilot's weight in front. Great ingenuity is also apparent in the location of the fuel tanks in the wing roots, and in the cooling arrangements for the engine. The airplane's building costs came to approximately $3,000—plus the usual thousand or two man-hours.

But that commitment of your time, I think, is a large part of the appeal of home-building. It is a way of becoming totally involved in flying.

181

9

Moving Up

The great majority of the world's more than one million pilots will- never advance beyond the private license that enables them to fly a single-engine landplane. They'll tell you that this is as far as time, money, and commitments to family or business will allow them to go.

Perhaps. But I can't help feeling they're confining themselves to a small and well-trodden corner of the flier's world.

There are an almost infinite number of ways a pilot may extend and refine his skills. The topmost professional levels, admittedly, demand dedication and qualifications not everyone is blessed with. Yet for the successful they can be marvelously rewarding both financially and in terms of personal attainment.

King of the civil flying community is the senior airline captain. (More of him in Chapter 11.) With four gold stripes on each sleeve and a sprinkle of gold on the visor of his cap, he is a distinguished figure looking well worth his $80,000 salary, plus expenses. There are lesser captains under him, staying fit and waiting, waiting, for their turn in the left-hand seat. His first officer probably holds a captain's rating, and his second officer is a pilot, too. Flight engineers and navigators most often are not, although a number have made the switch and taken pilot training.

All members of airline flight crews have one thing in common: on one particular day in history when an airline was hiring, each happened to be the best-qualified guy available. Their credentials may vary more widely than you'd think: from several thousand hours of multiengine turbine transport time (most likely in the military) down to not much more than a basic PPL for a lucky handful.

This is very often what your instructor of student pilots is- shooting for. There's little enough money in instructing, but it is a stepping stone, an opportunity to "build time"—accumulate hours aloft—with the hope that one day he will be tapped for a job with an airline. Or perhaps a business-flying operation or an air-taxi outfit. This is not to say that your instructor is not well qualified to teach you, or is uninterested in doing so. Most likely he has an abundance of youthful enthusiasm that com-

municates to his students, as well as the requisite skills to pass the stiff FAA tests for an instructor's rating. Still, he has a long way to go before he is more attractive to an airline than the surplus pilots furloughed during the last slump in the flying business or the hundreds of ex-military fliers with a thousand hours of F-4 time.

Some national airlines—British Airways, for one—take kids just out of school, with no flying experience at all, and put them through the whole business from scratch at the state's expense. In consequence, they have some copilots who are extremely youthful but have all the necessary flying ability.

It's really the money that makes airline flying so special as a career. "Otherwise we're just highly qualified bus drivers," the more honest may tell you.

The next most sought-after job in aviation is probably as a pilot in some big corporation's flight department. Flying executives around pays well, often almost as well as the airlines. The equipment can be pleasingly fancy: twin or even four-jet transports as well stacked with interesting black boxes as an airliner. Though corporations are vast, their flight departments usually are small, so there is none of the tiny-cog-in-a-huge-machine feeling that may afflict pilots working for an airline. You usually get home at night; and if you don't it may be because the boss has decided to stay over in some pleasant place like Nassau or Virgin Gorda. And you're right up there, mixing with the brass. You probably know your board chairman better than ninety-nine percent of his employees.

Then there are the little commuter airlines, back and forth all day between suburbia and the big smoke. And the air-taxi outfits, who may fly twin jets or piston twins. Jets by law must have two pilots, but light piston twins may be flown solo, and working on their own does appeal to some pilots. Crop-duster pilots are a special breed, rugged individualists doing a dangerous job, sometimes making a year's money in six months and then taking it easy. There are guys who ferry airplanes around the world for a living, and they too are an old breed, tending to be loners, I think. There are the sales and

184

demonstration pilots, well-dressed, always wearing the latest tie, showing you how easy the new model is to fly, and quick at doing sums to work out how low the monthly finance payments can actually be.

Few military pilots stay in the service all their working lives. Unlike the airlines, the military needs lots of young pilots but few middle-aged ones, so out they come after ten or fifteen years, with all that high-quality time. The very best become test pilots, and the very best test pilots become astronauts—for a bit, because astronauting tends to brief, bright moments of glory, and then you need a job again.

Test pilots usually have a master's in engineering, ferocious intelligence, and small, quick, neat hands. They are the superpilots, men who can fly anything, men who take the raw machine and mold its handling till it is civilized enough to keep less-skilled mortals out of trouble. Test-flying, particularly of military machinery, is still quite a dangerous occupation, though these days it is largely flying to measure the airplane's performance rather than pure handling. Still, I feel other pilots respect the test pilot for his courage as much as his superior flying ability.

For the holder of a PPL, however, there are less ambitious opportunities. He may learn to elaborate his technique by venturing into soaring, for example, or sport flying such as aerobatics or racing, or—if he is willing to put up with the book work and the grind of practice-makes-perfect—by earning an instrument rating.

Simply checking out on airplane types you've never flown before can stretch your mind and talent. Generally it involves no more than studying the owner's manual and flying an hour or so with someone well versed in the beast's ways. Some pilots become "type hogs," aiming to list in the "remarks" column of their logs as many different species as they can manage. Every new type teaches you something, even if it's only that you aren't quite as smart an aviator as you'd thought.

Taildraggers are particularly humbling for pilots raised on nosewheel landing gear, which means virtually everyone, because nosewheel airplanes are easier to handle. So is automatic transmission in an automobile, but it's awfully dull. Tailwheel air-

planes are the sports cars of flying, the antiques, racers, or aerobatic types that are sensitive and responsive to a skillful touch. Skillful but not superhuman; even a low-time, nosewheel-trained pilot can check out on one of these.

The first novelty you note, after settling into the seat, is the restricted forward horizon. Because the tail sits low, the view from the cockpit is of a bulbous airplane nose. Well, to be sure you don't take a propeller-sized munch out of anything hiding in your dead spot on takeoff, you just have to weave a bit while taxiing. You discover the next novelty as you're under way: the thing doesn't want to go straight. Because the center of gravity is behind the main wheels, instead of ahead of them as in a nosewheel airplane, a taildragger is inclined to amble. It's a little like reversing an articulated truck. You learn to be quick and deft on the rudder pedals, and also to taxi s-l-o-w-l-y, which increases control over the situation and reduces the inertia lurking in the machine should you somehow make a pig's ear of maneuvering it. You very soon pick up the knack of applying opposite rudder the moment a turn starts.

Most taildraggers have steerable tailwheels linked to the rudders by springs, and mainwheel brakes operated by pressing your toes on the top part of the rudder pedals. But to brake an old 65-horse J-3 Cub or one of its descendants, you have to press with your *heels*, which is not so easy. And since the J-3 is splendid fun to fly, as well as the cheapest way two people can get airborne under one roof, it is likely to be with us for years to come. Some taildraggers have brakes that come on only at the full extent of rudder-pedal travel; these require a special touch. Some have free-swiveling tailwheels, and need a lot of downwind pedal pressure when taxiing across-wind to stop them weathercocking into the wind. Some have swiveling tailwheels you can lock straight for takeoff—and you had better remember. The old P-51 Mustang had a tailwheel that locked if you held the stick back, but free-swiveled if you pushed it forward.

World War I types are definitely not for beginners. Mostly they have no brakes and fixed—unsteerable—tail skids. They go pretty straight if you hold the stick back; to get them to turn you have to

185

Two versions of 1930s Ryan Sport
Trainer (in-line engine below, radial opposite),
a taildragger that is tricky
to keep straight on landing, but is prized by
antique-airplane buffs nonetheless.

Learning to fly off and onto water is
a big step up in pilot's education—even taxiing at low
speed (above) requires mastery of exacting
new skills. Cessna Skywagon shown here (and opposite) flies
Long Island commuters to lower Manhattan.

1973 Reims/Cessna Pressurized Skymaster
is the only twin engine of its kind with centerline
thrust—if one engine loses power it
will continue to fly straight, requiring
only moderate skills to handle.

wind a short way, the best bet may be to close the throttle (a waterplane always tends to sit on the water with its nose into wind) and drift backward. (But remember to retract your water rudders; otherwise they will oppose your air rudder.) To go across-wind, apply just enough power to be able to turn the nose slightly out of wind, and let the wind sail you sideways. Of course, you try always to come up on a mooring buoy or the ramp into wind, as in a sailboat. To turn either way from into wind may require a great blast of power; if you've any distance to go, you can do it at speed "on the step." Turning out of wind, the wind and centrifugal force tend to balance each other. But take care any time you are taxiing downwind and begin to turn into wind. Here centrifugal force and the wind are in cahoots. Best throttle right back and turn only at minimum speed or you may dig a wing tip in or even capsize. "Step turns" need care, and your instructor will probably make you spend time practicing them.

Amphibious floats, or amphibious-hulled seaplanes, with built-in landing gear handle just like ordinary floats. If you should absent-mindedly land on land with the wheels retracted, no problem. You may not even do any damage. But if you land on water wheels-down you most likely will wreck the aircraft, for the wheels dig in and tip you over straight for the bottom. It is best to go through the whole rigmarole aloud: "I am about to land on *water* and therefore my wheels are UP, check UP." And so on.

Multiengine flying introduces the lightplane pilot to new complexities in aircraft systems and performance. It is flying of a fundamentally different kind. The disciplines of multiengine flying tend to foster a passion for precision. A single-engine pilot, for example, will just taxi away from the ramp without worrying about it; a multi man will taxi quite slowly, and with his nosewheel running exactly down the white-painted centerline. Having two engines does eliminate the need for the single-engine pilot's constant consciousness of a place to set down; it also introduces such new concepts as minimum control speed, best-rate and best-angle climb speeds, and fuel systems capable of cross-feeding. Your first flight in a twin may also be the first time

you encounter constant-speed propellers, let alone feathering ones.

Many singles can be more or less loaded up with people and baggage without raising problems for the pilot. Most twins cannot. Usually you must do some calculating when all the seats are filled or the baggage compartment is fully loaded. Fuel loading must also be taken into account. A center of gravity aft of the airplane's limit may make it unstable in pitch; and simple overloading of a twin may destroy its ability to climb on one engine, should you ever have to.

Because the effectiveness of the rudder in counteracting yaw varies with your airspeed, whereas power is pretty constant, there is always a minimum speed below which you just can't keep straight should you lose power on one engine while you've still got full power on the other. And you want one speed to get the best rate of climb, and another for best angle, if there is rising ground ahead—and these speeds are different when you're flying on one engine. So the twin pilot flies with these speeds much in his mind, especially on the climb-out and on the approach. Losing an engine just after takeoff is still a real emergency in a heavily loaded twin. You've quickly got to identify which engine it is (not easy, and best done through your feet: "Strong foot, good engine, weak foot, dead engine"), and get that propeller feathered to reduce its enormous drag while it is still windmilling; get the gear and flaps up to further reduce overall drag, hold best climb airspeed; tell the tower you're coming around for an immediate return and landing; fuel and switches for the dead engine to *off*; watch your airspeed; stay on top of things. Above all, in the words of my old instructor, "Fly the airplane. Everything else is secondary." Losing an engine on takeoff doesn't happen often, but it *does* happen, first power reduction being a traditional time for engines to let go if they're going to. Consequently, I was taught never to be in too much of a hurry to come back from takeoff power to climb power.

The modest complexities of twins quickly become a source of real pleasure: getting engine RPMs synchronized till their sound is a smooth hum with no "beats"; attending to cowl flap positioning for en-

192

gine cooling; setting mixtures and fuel flow with the exhaust-gas temperature gauges; holding best-rate speed within a knot or two; going early from mains to aux tanks to be sure they're feeding; and so on. I love the feeling of power being so close: those big bulbous engine cowls on either side of you, just outside the cabin windows, and the whirling propellers so near. I love running my eye and finger down the compartmentalized check list, where every action is in its place: preflight, start-up, taxi, takeoff, climb, cruise, descent, landing, after-landing. "It's a matter of housekeeping," my instructor used to explain. "Everything tidy, everything attended to." There's nearly always something to attend to, something to think about. Why, for example, does one fuel-flow meter always seem to be showing half a gallon an hour more than the other? It always seems to me to resemble cooking a fancy meal for guests, particularly when you're on an instrument flight plan. It's not enough to be just doing what you're doing. You must be checking what you've just done and thinking ahead to what you will need to be doing next, all at the same time.

And while you can land a twin like a single, whopping the flaps down in one go and horsing it around from base onto short final, somehow that doesn't seem to suit the airplane. More rewarding is

to make all changes in configuration, power, and airspeed as smooth and imperceptible as possible: gently back to seventeen inches on downwind, a smidgen of flap, a smidgen of trim, drop the gear as airspeed bleeds off, a big-bomber turn onto base, still carrying seventeen inches, full flap early and a long straight final approach, propeller levers fully forward and throttles closed, and grease it on—all done so smoothly that the passengers don't notice a single thing until the wheels touch. I never fly even the smallest twin without a feeling of kinship with those crafty old aviators who flew the Connies and DC-6s and 7s before the airlines went to jets. If it was good enough for them simply to aim to create the smoothest possible ride for the folks in back, me too.

The simple multiengine rating on your license entitles you to fly any light twin up to 12,500 pounds (all-inclusive) weight. For bigger machinery and for jets, you must also have a rating for the particular airplane type you want to fly, which means another written exam and flight test. (Bigger airplanes also are required to have at least two pilots.) Type ratings naturally concern the professional rather than the private pilot. For a business jet, a week's ground school is required, some simulator time, and then perhaps a dozen hours in the real machine. You fly with an instructor even if you've

193

Fliers, like car buyers, flock to air shows to see what's new—and what represents the next step up in flier's world. Gaggle of aircraft at annual Reading, Pa., show (l) is numerous and varied. King Air's twin-engine, 14-passenger executive carrier (below l) is one direction for the aspiring pilot to go. The helicopter under study by the quartet below is another.

Aircraft for Business: Ten-passenger Rockwell Sabreliner (below) is descendant of F-86 Sabrejet fighter used in Korean War. Maximum cruise is 536 mph (862 km). Two Rolls Royce Spey turbofan engines give Grumman Gulfstream II (far l) a max cruise of 588 mph (946 km). It was first bizjet with true transoceanic range, is one of the biggest (up to 19 passengers) and most expensive. Gulfstream I (l) is earlier turboprop model.

Two types of Gates Learjet: G-BBEE is Model 25B, with pure jet engines. N362GL is newer Model 36 with more economical fan jets. Lear is most popular business jet, has outstanding mile-a-minute rate of climb and a 500+ mph cruise.

moved to show the maximum allowable Mach or airspeed, whichever was lower—it worked these out for you. It had, it had ... what it didn't have, we soon discovered, was an autopilot that was yet fixed. But you didn't hear me complain, I promise.

I made a mild hash of trying to land the bird, though. "Aim short, aim to touch down before the threshold," he told me, but even so I came in too high and also managed to get the trim wound too far back. "I have it," he said firmly, and took control and salvaged the approach for me. He was kind enough to say no more about it.

To an ordinary piston pilot, everything in a jet seems to happen with a rush. You rush down the runway for a vast distance before you fly. You continue to accelerate thereafter, then rush up to altitude quick as a flash, for fuel consumption only settles down to something halfway reasonable when you get high. At the other end, you rush downhill, get established on the glide slope as soon as ATC will let you, and hurry to touch-down.

The vital importance of these speeds is soon evident. On takeoff you quickly reach a speed beyond which you no longer have the option of not taking off; you *must* fly, since it's impossible to stop in the distance remaining. You lift the nose at another precise speed, climb at exactly a third, come back to a fourth for noise-abatement reasons, resume your climbing speed until your climb Mach is reached, and then proceed at exactly that rate. And so on. You even have little movable bugs on your airspeed/Mach indicator which you can set to mark critical numbers.

And the comfort of it all is impressive. There is no vibration and almost no engine noise in the cockpit, just the smooth rush of air around it. I recall one captain who, when his airline first went to jets, said he didn't want to fly them. "I'll sit this one out," he said. But then they took him down to the hangar and showed him one. "When I took a good look at that airplane the hair stood up on the back of my neck. And I said, 'When does the next class begin?' After the first few hours of flying the thing, I just wanted to stand around and look at it. It had been such a thrill that I got a lump in my throat."

Exactly.

201

10

Aerial refueling (above) vastly extends range of F-111, but even without this boost plane enjoys transcontinental range. More limiting to continuous flight is fatigue of pilot anchored in not-too-comfortable seat. Practice landing (l) may be complicated with simulated mechanical failure or other emergency. 207

from pilot training somewhere along the way. Those who turned out to have some intractable problem, such as persistent and irremediable airsickness, might have become ground officers or might have chosen to return to civilian life. Those who, despite high intelligence and keenness, turned out to lack some aspect of the curious assembly of qualities and aptitudes required of Air Force pilots might have gone to swell the ranks of the navigators, whose work these days includes much more than merely finding the way to the destination. Indeed, the navigator in an F-111 is called a "weapons system officer," WSO for short, pronounced "wizzo."

A pilot is typically twenty-four or twenty-five years old when he first joins a squadron; a navigator may be a mite younger. He can expect to remain in that squadron and based in one location (such as Heyford) for three years. Promotion to captain may be expected four years after joining the Air Force. Promotion to all ranks is by merit; but a pilot might expect to make major within ten years. Higher rank thereafter will be increasingly difficult to attain. Not every Indian can make chief. Those who join the Air Force from the USAF Academy, and some of those entering through the other two programs, are regular officers, in the service for an indefinite term, up to thirty years or longer. The others are reserve officers, in for five years, though they can stay longer (up to twenty years for retirement) and may indeed be offered a regular commission. But it is the reserve officers who go first when there is an RIF program—reduction in force. Most pilots leaving the Air Force try to join an airline and hope to be leaving at a time when the airlines are hiring, which is not always, by any means. Not so many of the navigators will stay in flying when they leave. Airlines no longer carry navigators except on transoceanic runs, and few are needed. Most navigators, when their time in the service is up, expect to go into business, or out into the "normal job market," whatever that is.

Navigators may, after three years as a navigator with the Air Force, apply for pilot training. Those who do include some of those spectacles-wearers who could not originally meet the strict eyesight requirements for pilot, and even a few of those who

208

Aardvark at Rest: So called because of anteater-style nose, 40-ton F-111 is more complicated to fly than 747, has some 425 buttons, knobs, and switches in cockpit. Despite F designation, it is a bomber with fighter capabilities on the side.

Below: General Dynamics F-16 is small, lightweight, remarkably maneuverable, relatively low-cost fighter. Bottom: Grumman F-14 Tomcat is multipurpose, carrier-based U.S. Navy fighter with variable-sweep wings, like F-111. Left: Lockheed C5A Galaxy is colossal USAF freighter that can carry two M-60 tanks or 16 three-quarter-ton trucks.

McDonnell Douglas F-15 Eagle
(below & bottom) has Mach 2.5 speed combined
with exceptional maneuverability.
It has climbed to 65,500 ft (20,000 m)
in two minutes. U.S. Navy's F-14 Tomcat (l)
has variable-sweep wings, part-titanium
construction, versatile performance.

213

Blackbird: SR-71 is a strategic photo and electronic reconnaissance aircraft that can fly at more than Mach 3 and higher than 80,000 ft (24,384 m). It has flown New York-London in under two hours. Nickname derives from heat-dissipating dark matte finish.

earlier washed out of pilot training. It is felt that the experience they have since gained may compensate for any initial disability.

The Air Force, I was told at Heyford, is most interested in making every officer a potential manager and runs several schools to this end. Many of the junior officers at Heyford were pursuing further education in their off-duty time, perhaps studying for a master's degree in business administration.

Thirty days' leave a year is standard for the pilots at Heyford, but they must pay their own way back to the U.S. if that is where they wish to go. They therefore often take their leaves in Europe, going skiing in Switzerland in winter, to Greece in summer, driving around continental Europe or Britain, or maybe showing relatives from the States around their new home, England. Weekends are usually free, there being an unwritten but longstanding agreement between East and West that nobody will start anything between Friday and Monday. Weekends may be spent playing golf or tennis, soaring, sightseeing, or visiting London, which is less than two hours away. And, as fighter pilots have done since aerial warfare began, the men at Heyford have parties. It is a nice life.

The F-111 is not my idea of a fighter. It may weigh forty tons at takeoff, and it is more complicated to fly than a Boeing 747. In truth, it is a bomber with combat capabilities on the side. It was the first service airplane the pilot could redesign in flight by changing the wing sweep. For takeoff and landing, the wing is set forward, a Venetian blind of high-lift slat and flap and spoiler surfaces. In flight they are fully aft, forming a paper-dart platform that slips most easily through the air at twice the speed of sound. The USAF has never named the F-111, but the crews have: they call it the Aardvark, after its anteaterlike proboscis. I do not think that anyone knows what its maximum speed is, for it is not limited by available power but by the temperature of speed. It is an aluminum airplane, and above Mach 2.5 its skin begins to attain temperatures (from air friction) at which it can lose its strength. This is thoughtfully taken care of, like everything else in the machine, by a warning device. When criti-

cal skin temperatures are approached, a light goes on and a digital readout starts a countdown from 300 seconds. Thereafter a "total temperature" warning illuminates the panel, and by then you had better have pulled the power back and slowed down.

The F-111 will do about Mach 1.2 "on the deck," and because of its size it kicks up an enormous shock wave. Low-level supersonic bombing runs are practiced at a range in an empty part of Nevada. Wrote one visiting F-111 pilot: "I have seen the entire valley floor erupt into a cloud of boiling dust, as if struck by the flat of a giant hand, following a low-level supersonic bomb run."

The F-111 has extraordinary range. Its variable-sweep wing enables it to take off with a huge load. The center part of each wing is filled with fuel out to the tips; there are huge tanks in the fuselage behind the crew compartment; there are saddle tanks around and on top of the engines; there are tanks in the weapons bay; as many as six external tanks may be carried on pylons under the wings; there is even a tank in the vertical stabilizer, though this is not filled on the ground, but left empty for fuel from the other tanks to expand into. Even the structural heart of the airplane, the wing box that joins the two hinge pins for the swiveling wings, is hollow and can double as a fuel tank. (Ten electric fuel pumps are required to move all this kerosene.) On internal fuel alone an F-111 can fly for some 3,000 miles (4,827 km). It can cross the Atlantic without needing a tanker top-up, or fly coast to coast and have enough left to shoot several touch-and-goes on arrival. The range derives not just from the fuel capacity, but from the cruise efficiency of the turbofan engines.

The most unusual feature of the F-111 is its TFR—automatic terrain-following radar. TFR is no novelty in military aircraft, but an *automatic* one is. The crew can select a wide variety of altitudes for the TFR to follow, down to as low as 200 or 250 feet. In addition, a three-position ride-control switch gives the options of soft, medium, or hard ride, enabling the crew to select how closely it wants the machine to follow undulations in the terrain. "The Cadillac of fighters," they call the Aardvark. The

Lockheed F-104 Starfighter (top) is a single-engine,
single-seat fighter designed by Clarence "Kelly" Johnson,
who also did Shooting Star, U-2 spy plane, and SR-71.
French F1-E Mirage (bottom) was unsuccessful contender
as replacement for 104. Most of Europe took F-16.

airplane's flight-control system employs automatic stability augmentation: triple pitch, roll, and yaw circuits, operated by a computer that commands corrections through the aircraft's controls far more quickly than a human pilot could, work constantly to compensate for any deviations in the aircraft's motion due to gusts or turbulence. The result is a strangely solid, steady ride at low levels and high speeds. When a hard ride is selected, the airplane delays its pull-ups over rising ground, and pushes over the tops of hills at up to zero g. Nor do you have to fly in a straight line when using the TFR. You can fly around peaks rather than over them— without ever "seeing" them with your eyes.

To experience the F-111's TFR in action is uncanny. Ground checks of the system, its radars, and its fail-safe devices are made before takeoff, and further checks are performed once the plane is airborne and cruising at altitude. On the pilot's command the aircraft begins descending at some 10,000 feet per minute. At 5,000 feet (1,524 m) the radar altimeter identifies the earth's surface, and the TFR "acquires" its target, the ground. Its first action is a nerve-racking one: it instructs the autopilot to pitch down into a steeper descent! But a smooth level-off is made once the programed height above the terrain is reached. Perhaps the most extraordinary part of the whole performance is that the pilot's control stick does not move. The TFR's computer sends its climb and dive commands directly to the flight-control system, by-passing the pilot's controls. The crew monitors what the automatics are doing, but it could do so with arms crossed if it wished—except that the pilot may need to apply power if there is a particularly steep mountain ahead. Mountains at night slide by as half-seen hulks. In cloud by day they may only be distinguished as a darkening of the mist to one side as they approach and recede.

The WSO watches his own radar, a big ARS (attack-radar system), which continuously sweeps through a whole band of frequencies so that it is nearly impossible to jam. By just punching a button he can summon up from the computer preset target coordinates, and cross hairs on his radar screen, deriving their information from the inertial navigation system, reach out and settle on the target. More automatic systems determine the ideal moment for bomb release.

What if something "goes wrong"? The vital systems are all duplicated or triplicated, with surviving elements taking over automatically should one fail. And all features of the TFR and control systems are elaborately checked out before they are used. If some drastic malfunction should still occur, fail-safe circuits command an immediate and shuddering pull-up away from the ground.

Should something happen so disastrous that the aircraft must be abandoned, the crew fires not ejector seats, as in most military aircraft, but the entire cabin, like a tiny spacecraft, on twin rockets that quickly raise it far above the carcass of the airplane. In due course a colossal parachute, like those of the Apollo spacecraft, automatically deploys and lowers the cabin module gently to earth. The descent to earth from high altitude may take as long as ten minutes. The crew may release metal chaff on the way down so it can be tracked by ground radar, and once below 15,000 feet (4,572 m) it can open the cabin windows for some fresh air. An air bag inflates to cushion touch-down on land; others keep the module afloat in water. In mountainous terrain the crew is advised to close the windows before touch-down in case the cabin rolls down a slope. Should the module touch down at sea when a swell is running, and start to take on water, the pilot's control stick may be connected to a bilge pump and moved backward and forward to drain the cabin and pump up the flotation bags simultaneously. All the while the crew can talk to rescuers by radio. Baldly described thus, the ejection ritual seems almost unreal, but in the dozen or so times it has had to be used, it has worked well. One F-111 pilot who has used the ejection system told me he had suffered no ill effects, except that he is now slightly shorter, owing to the acceleration forces and consequent compression of his spine when the rocket fired.

F-111s proved to be potent weapons during the Vietnam conflict. They flew a great many raids —3,500 missions—either operating singly or acting as radar "pathfinders" for less well-equipped fighter-bombers of other types. Only six F-111s were lost, which is a far lower casualty rate than was general

217

Anglo-French SEPECAT Jaguar is
both an advanced operational trainer and
a high-performance tactical and recon plane.
It flies for oil kingdom of Oman, as well
as for RAF and French Air Force.

in that unhappy war, and this despite the fact that they were operating against the most highly defended targets in air warfare's history: rail yards, airfields, and SAM (surface-to-air missile) sites in the small flat valleys around Hanoi and Haiphong, the heart of North Vietnam. Despite the aircraft's inherent electronic complexity, the mission abort rate due to malfunctions was less than one percent.

By choice, the F-111s flew at night in vile weather. They would start their two-and-a-half-hour missions in a high-level cruise from their base in Thailand. Before reaching enemy territory they let down into the monsoon clouds till they were skimming the jungle a mere 200 or 250 feet up (61–76 m). At the start of the letdown they would be in moonlight, with jagged peaks poking up from the cloud below. Then the moonlight would fade into the mist and the night grow even darker around them as invisible peaks shot past. "I won't say I wasn't worried!" said one WSO afterward. "We were always nervous, no matter where we were targeted, because flying as low and as fast as we did is inherently dangerous. At 500 mph you are only a quarter of a second from hitting the ground if anything goes wrong. Think about flying around in daylight and good weather only 200 feet above the ground and going up and down over hills and valleys maintaining this height." And, indeed, they posted a note on the bulletin board in the officers' club to the effect that while the SAMs were no better than fifteen percent effective, and the Triple A (antiaircraft artillery) less than five percent, hitting the ground was invariably a hundred percent deadly.

The pilots proved the ability of their gadgetry to find the airplane's way to tiny, sometimes unseen targets, which often made their arrival in target areas a total surprise to the defenders. This was achieved, too, in spite of North Vietnamese radar coverage at a level the Americans described as "unbelievable." "There was no such thing as coming in underneath it in the Red River Delta," said one F-111 pilot, quoted in *Air Force* magazine. "The place was so small, so flat, and so heavily defended that they were looking in all directions all the time for attacks. And their gunners had had more practice in the previous five years than any gun or missile

219

Hawker Harrier (below) is vertical-takeoff
jet fighter serving RAF and U.S. Marine Corps.
Four thrust nozzles are turned down for
lift-off, back for flight. Right: British
Phantom FGR Mk 2 on training mission.

crews in history."

Like the human eye, radar is a line-of-sight device. It cannot "see" through hills. So the F-111s,
flying by TFR at minimum altitudes, were invisible
to the enemy until they crossed the last hills before
the defended central plain. Pilots likened the motion
of riding the F-111 low down by TFR to skiing. And
once they came skiing down the slopes of the mountains around Hanoi their electronic gear would an-
nounce that enemy radar pulses were reaching
them. Even so, the defenders seemed unable to track
the F-111s accurately, so low did they fly.

One F-111 crew on one of the earliest sorties
found its arrival over Hanoi so unexpected that
every neon sign and streetlight was still on, and so,
to their disbelief, were the runway lights at the airfield they had come to bomb. As they flew on, sections of the towns below started going black as the
North Vietnamese belatedly realized what was hap-
pening. On a night with a 300-foot cloudbase they
had never expected an attack.

The F-111's principal targets were the SAM
sites that were proving a menace to other aircraft
types bombing from higher levels. Here is a measure
of their effectiveness: while approximately one hundred SAMs were launched every night at the beginning of the F-111's campaign, by the end often not
one would be fired in a whole night. Near the end of
the war the U.S. strike planners could dispatch just
a single F-111 to a target with near certainty that
the mission would be effective.

At Heyford in England today the crews do not
fly every day, but perhaps a dozen times a month. A
training mission of two or three hours' duration,
with its methodical preparation and debriefing, just
about fills a working day. Crews report for work at
7:30 A.M. and start their flight planning: two or
three hours studying weather forecasts and determin-
ing the flight route, which depends on whether there
is to be bombing on ranges off Britain's coasts, bal-
listics bombing practice with its involved computa-
tions, or simulated bombing judged by radar scoring. Guns are not fired. One of the policy decisions
that came out of Vietnam was that it was not worth
risking a $15,000,000 airplane to strafe a $10,000
truck. But they do drop practice bombs, which re-

220

SAAB Viggen is Swedish Air Force
multipurpose fighter designed to operate
off short stretches of highway in countryside should
more vulnerable airfields be
bombed and made unusable in wartime.

lease smoke markers on impact with the sea. Accuracy is scored by triangulation by the range crews. Level and dive bombing, and low-level weapon delivery, are practiced from medium altitudes. Bombing can also be practiced without releasing any bombs —by employing a camera attached to the WSO's radarscope which records the moment of "delivery" for later analysis as to accuracy, or by transmitting a radio tone which ceases the moment a "weapon" is released (here ground radar at the range is used to analyze accuracy). The ranges around Britain that the Heyford crews use are all RAF—as indeed, technically, is their own base at Upper Heyford—for there is very close cooperation between the air forces of the United Kingdom and the United States.

Flight planning completed, the crew goes out to its aircraft, receives a report from the crew chief on its serviceability, and begins the elaborate and careful checks of its functioning. These continue af-ter engine start-up and throughout the flight.

A ladder is needed to board the aircraft. It is entered not by the one-piece canopy usual on fighters, but through two side doors. Visibility from the cockpit (except forward) is poor by fighter standards. This is perhaps not terribly important, since in the F-111's unique environment—alone, at night, and right on the deck—it is not likely that any enemy fighter will find and engage it.

Immediately in front of the pilot are the two big round dials of the flight-director system, with banks of vertical-scale flight instruments flanking it and a battery of warning-light panels below. Engine instruments form a panel on the right, between the pilot and WSO—so far, nothing unfamiliar to any pilot of turbine airplanes, except that the vertical-scale instruments are peculiar to the military. In front of the WSO is the huge attack radarscope, with the bomb-nav control panel below it. Up top and

centered between the two crew members are the TFR scope and the homing and warning radarscope. Ordinary radios are mounted below these, as are stand-by flight instruments should the main flight director fail. Both occupants have stick and rudder controls; WSOs may try their hand at flying the airplane, even though they have received no formal training as pilots. (Instructor pilots also ride in this right-hand seat when checking other pilots.) Throttles, wing sweep, gear and flap levers are on the pilot's left. If there is one thing to surprise a civilian pilot and bring home the complexity of the F-111's systems, it is the number of knobs and switches in the cockpit. There are about 425 of them scattered across the panel, between the crew, along the cockpit walls, and even up and behind them. The windshield is perhaps the closest the F-111 comes to having an Achilles heel: it can occasionally be shattered by collision at speed with large birds. A new and

stronger one is now being fitted.

The airplane is taxied quite slowly. The crew sits right above the nosewheel. The two main wheels are close together and well behind. During a tight turn on the ground there is a feeling of teetering. Acceleration for takeoff is quite fast, despite the aircraft's weight. Nosewheel lift-off is made at around 142 knots, and the pilot immediately checks forward on the stick to prevent overrotation. He must wait a moment before selecting gear up, since the wheels are automatically braked at the start of the retraction cycle, and were they still to be in contact with the runway, both main tires would burst.

As speed builds up the pilot moves the wings back. The F-111 automatically compensates for the trim change, and the four inboard wing pylons also rotate as the wing sweeps to remain streamlined with the airflow. Most speeds may be flown with a wide range of wing-sweep angles. As a trick for visi-

223

tors' days at the base, F-111 squadrons sometimes fly a two-ship formation that starts with the leader flying wings-forward and the slot man wings-back; they then reverse the sweep angles as they go by.

From Heyford an F-111 might climb to 28,000 feet (8,534 m) and proceed to Land's End, the westernmost tip of England, and there let down to around 1,000 feet (305 m) for a low-level run at 450 knots north across wild moorland country, using the TFR, across the Bristol Channel, and through the Welsh mountains. At Liverpool it might cut out over the Irish Sea, swinging back inland and through the hills of the Lake District before turning south and working the bombing range at Jurby, on the Isle of Man, for a half hour. After dropping six practice bombs, the F-111 might climb back up to 28,000 feet and swing east to do some radar-bomb scoring for twenty minutes off the east coast of England, then do a couple of runs at 5,000 feet (1,524 m). If the weather is bad here, there is always the Tain range in the north of Scotland. In a single sortie the Heyford crews often see more of Britain than many Britons do in a lifetime. Their flying over Britain, for peacetime safety reasons, is conducted under fair-weather conditions. For one thing, low-level military routes over Britain are not published, so that civilian fliers do not know how to keep clear of them.

Bombing practice completed, the F-111 might shoot some practice landing approaches of various kinds at several military bases on the way home; besides ILS, PAR (precision-approach radar), and the usual letdown aids, the aircraft's own radars can be programed to simulate an ILS descent into any air base. Then back to 21,000 feet (6,401 m) for a standard military TACAN penetration return to Heyford, where enough fuel will normally be left for some touch-and-goes, perhaps with some minor emergency simulated, such as an engine out or flap failure. After landing, a maintenance debrief with the crew chief, then back to the squadron for an hour-long debrief on the mission. Then home, for a single such mission, with its preparation and debriefing, fills a long working day.

"We also get tasked for a lot of exercise flying," one of the Heyford pilots told me. "Antiship warfare, in cooperation with NATO navy ships; and simulating close air support to the U.S. Army on exercises in Germany. But mostly F-111 flying is a very stable type life; those horrendous two- and three-month TDYs [temporary duty away from home base] are not so common now. Though the F-4 pilots pull a bit more TDY than we do, from their own bases in Spain and Germany. British weather does make our training more difficult. New pilots arriving here require a special check-out on it. We do go down to Italy sometimes to work one of their ranges, and stay overnight or over a weekend."

If there is one thing that makes an F-111 pilot different from other fighter pilots, he told me, it is that "a good F-111 pilot is a fantastic instrument pilot." But for all the awesome responsibilities of being part of the West's nuclear deterrent, and the sheer work involved in training in such a complex machine, they are still a bunch of guys very much in the fighter-pilot tradition, which is a light-hearted one. One of their squadron crew rooms at Heyford was painted in camouflage colors to match their airplanes. Odd mementoes of their travels decorated the walls, as did a well- (but accurately) used dartboard, a stuffed aardvark in a glass case, signs announcing that "Fighter Pilots Do It Better," and an honors board boldly labeled "The Xaviera Hollander Happy Hooker Award." (On it were emblazoned the names of three squadron unfortunates who had contrived to drop [accidentally] the arrester hook that the F-111 carries for emergencies such as brake failure.) Their parties can be mildly riotous. At one they had a whole barrel of Rhine wine, brought back from an exercise in Germany, and also the entire staff (in their abbreviated uniforms) of "bunnies" from a London night club. Fighter pilots seem to be a special breed, much the same in all air forces at all times. But they are a rarer breed now that their machinery comes so dear, and possessed of enormously more complex skills than their predecessors.

I like the story of the USAF ops officer in Thailand who asked the first F-111 pilot who arrived there: "Do you people have smart [i.e., guided] bombs?" "No," he was told, "but we've certainly got smart airplanes." And the North Vietnamese, who never did learn to detect intruding F-111s, had their own nickname for them: "Whispering Death."

Wide-bodied 747 is the biggest commercial airplane ever built and, after billions of miles flown, demonstrably the safest. Despite takeoff weight of 775,000 pounds (351,540 kg)—two tons of fuel are consumed taxiing to runway—it has lighter controls than Old Faithful 707, is easy to fly. Left: Ponderous takeoff. Below: Complex instrument panels, deluxe dining in First Class.

ceivable that anyone could learn them all, but they are grouped and positioned in ways common to all large aircraft, and so are quickly familiar to the crew. Many are needed only infrequently (the entire engine starting panel but once a flight, for instance), and some, such as fire extinguishers, are only there in case. Even so, it takes an experienced man a week or two of ground school to learn the systems of a new type of airliner.

Almost every action on the flight deck proceeds according to the litany of the check list, with one man reading out the checks and actions while the others perform them, calling out "Checked" or "On" or "Off" or "Within limits" or whatever. They know the procedures perfectly, but they always go through the check lists aloud, so there is no possibility of anything being omitted. These check lists are less interminable than those that got the Apollo astronauts to the moon and back, but they are still elaborate.

That the engines are running is known to the crew mainly from the indications of their gauges, for they can barely be heard in the cockpit. Jet engines run hottest at the moment of ignition during the starting cycle, before they are turning fast enough to cool themselves. The huge Pratt and Whitney turbofans in the 747 must be watched carefully, for they run very hot and are prone to "overtemping" during starting and when taxiing downwind. These engines are the 747's *raison d'être*: the airplane was first conceived when engineers realized that very large, hot-running turbofans in a huge aircraft would be highly fuel-efficient, enabling substantially lower passenger seat-mile costs.

The 747 must be taxied slowly. It is difficult for the pilots to gauge their speed from their high cockpit, so they monitor the groundspeed readout of the INS—surely the world's most expensive speedo. While the captain taxis the plane, steering the nose-wheels through a small tiller, his first officer takes down their flight clearance, and the flight engineer computes the takeoff data—speeds, distance of run, and stabilizer setting—taking into account the weight of the plane, the air temperature, and so on. Every extra degree of air temperature means another fifty feet of takeoff run, and every tenth-of-an-

inch drop in barometric pressure means another hundred feet. As the plane trundles slowly across the airport, the crew can look down on the older jets they pass and even onto the roofs of the terminal wings and into the control-tower cab.

After waiting its turn in a line of jets, our 747 is cleared to go. She has already burned a ton of fuel, just taxiing. The captain moves forward the power levers to seventy percent, still holding the brakes while he checks the sixteen engine gauges for a moment. Then brakes off, and full power. Now the engines can really be felt shaking the airplane, and heard as a deep organ-pipe buzz. The acceleration is tremendous, but so is the speed required to fly. The takeoff roll seems to go on forever.

About forty-five seconds after brakes off, the first officer calls, "V one," by which moment the runway is fairly hurtling by beneath; then "Rotate," and the captain firmly lifts the nose first about 10 degrees, then to 15–17 degrees. "V two," and there is a decisive double thump from far behind as the main landing-gear assemblies, now off the runway, bottom against their stops. And the sensation of thundering down a highway at 200 mph in a 350-ton vehicle is replaced by the cushiony ride of flight. The captain orders, "Gear up," and the first officer moves a small lever with a symbolic wheel on the end, and panels open under the airplane and the huge landing-gear assemblies, all five of them, start to fold up into the belly. The panels close behind them—it is all automatic, once initiated by moving that small lever.

At 1,500 feet (457 m) flap retraction is begun, the nose is lowered, and power is reduced for noise-abatement reasons. (At only a few of the world's airports, those situated at the water's edge or out in the desert, is uninterrupted full-power climb to altitude permitted.) By now the captain has switched in the autopilot, and from here on he will direct the flight by feeding it instructions rather than by manual manipulation of the controls. "You don't fly this plane," he'll tell you, "so much as watch it."

If our takeoff seemed a moment of real drama, I think it is. Medical studies have shown that pilots' pulse rates are substantially elevated during takeoff.

233

It is the sole part of the flight that must be done in a fierce hurry, and there is always the anomaly that for a few precarious moments after V_1 the airplane, though still on the ground, *must* fly, since it can no longer stop safely.

Full power is resumed, and needed, for the airplane is still extremely heavy and climbing relatively slowly—it will take some forty-five minutes to reach its maximum initial altitude of 33,000 feet (10,058 m). A 747 can fly at 45,000 feet (13,716 m), but it cannot get there until it has burned off much of its fuel load.

By now all the tension of takeoff is gone, and the atmosphere has become relaxed—much more so than it ever got in the days when the captain was a junior copilot. Like most captains today, he prefers to run a fairly informal flight deck, for discord in the cockpit can put a damper on a whole trip. "I've walked onto flight decks when I was deadheading," he says, "when you could feel the tension between them like a wall." So he is easygoing with the crew, and they respond in kind. But his authority, however understated, is very real, and never remotely questioned by anyone.

As far as Gander the route is a simple matter of going from one beacon to the next. We are racing in the opposite direction to the sun, so that it seems to set with suddenness. The stewardess begins serving dinner to the flight-deck crew, bringing the trays in one at a time. She uses a coded knock on the cabin door; there is a code, too, for her bell downstairs, which she uses if there is trouble in the cabin that she or the purser thinks requires the captain's help—a drunk, perhaps, or a fight. It has happened.

A break in the clouds below reveals snow-covered islands and peninsulas gleaming faintly in the moonlight six and a half miles below. Lights glitter around a curved shoreline so that it resembles the open phosphorescent mouth of a deep-sea fish.

Before we reach Gander, Air Traffic assigns our altitude and Mach number for the ocean crossing. Separation is by time and altitude across those parts of the sky out of the range of ground radar; the other planes around us, at different altitudes, are invisible in the night but may be seen on our own radar. One hundred and ninety-two miles east

234

In lineup for takeoff, the 350-ton 747 dwarfs
its companions. It accelerates swiftly but requires a long
run, becomes airborne at about 200 mph (322 km).
Lifting great weight of this plane to 33,000-ft (10,000-m)
cruising altitude takes some 45 minutes.

Tupolev Tu-144 is Soviet Union's SST. Slightly bigger and faster than Concorde (Mach 2.35 cruise, 4,000-mile payload range), it carries flight crew of three and 121 passengers. "Ears" are retractable foreplane to trim craft for takeoff and landing.

of Gander the last indications from its radio beacons fade. Now begin the quiet hours of the flight, with little for the crew to concern themselves with beyond computing how much of their precious fuel will remain on arrival. They do not even have to listen constantly to the radio, as they do over land, for they are now using HF rather than VHF for communications, and it has a device called SELCAL that works like a hospital doctor's beeper: a bell and a flashing light will alert us should Gander or Shanwick Oceanic Control wish to speak to us. HF has quite a different quality from VHF. Instead of the latter's empty clarity, there is a heavy background of weird, unearthly whistlings and rustlings that rises and falls in strength and is loaded with high-speed Morse and a half-heard babel of other airplanes' position reports. There is also a noise like potatoes frying, and the strange warble of the SELCAL codes. But unlike VHF, which goes straight out into distant space rather than curving back to follow the horizon, HF travels around the world, reflected back from ionized layers of the upper atmosphere.

So the flight-deck crew, being unbusy, talk. About their families, and how the kids are doing in school, about anything—taxes, swimming pools, home winemaking. Parachuting, and how it has never appealed to them. The bargains you can get in London with the pound so low against the dollar. Sailing, or soaring. Which breed of dog, or handling Volkswagens in the rain. *Anything.* "I had a navigator once," says the captain, "who was new, and he said to me, 'I'm very sorry, but I think we're fifty miles off course.' And we were." And later: "When I started flying, the greatest danger was starving to death. Now it's driving on the expressway that scares me."

The captain perhaps started line flying on DC-3s before the war, when flying jobs were hard to get, as they often have been since. He got less than three thousand dollars a year then; now it's seventy or eighty thousand. But he doesn't talk about that. The other crew members aboard this plane, particularly the cabin crew, are not nearly so well rewarded. Is not the ALPA (Air Line Pilots Association) the most successful union in the history of labor relations? But consider the other side of the coin: in

237

Concorde: Cruise speed of this supersonic transport—1,450 mph (2,333 km) at 55,000 ft (16,764 m)—is faster than maximum of many military aircraft. A joint British-French effort bearing insignia of both countries, it carries crew of three and uneconomic total of 128 passengers.

what other profession do you have to put your job on the line four times a year, with proficiency checks and physicals? Indeed, only about one in ten of the young men who started flying with the captain in 1940 are still flying today. Flying wasn't so safe then; a commercial pilot's life expectancy was only about ten years before World War II (and much less than that during the war, even for transport flying—remember the famous "Hump"?), whereas now the captain can get life insurance at the same rates as anyone else. But his loss-of-license insurance is not so cheap.

Perhaps nobody really expected jet planes to prove as safe as they have. There were too many crashes in the early days, before operation techniques and training were so thoroughly worked out.

Concorde is probably a transitional airplane. It has too many inefficiencies to be profitable, enough environmental impact to be controversial. But its lovely speed inevitably points the way of the future.

Now their safety record is almost incredible; and the 747 has proved to be easily the safest passenger vehicle ever, despite all those misgivings about so many people in one plane. Few workmen can have seen the tools of their trade so drastically improved in their own lifetime as the airline captain—from the 21-seat, 180-mph, unpressurized DC-3 to the 365-seat, 650-mph, above-the-weather 747. Ice used to be a constant problem in the old DC-3s. They cruised mostly at between 4,000 and 8,000 feet (1,219–2,438 m), just where the icing was worst. The Boeing jets today all have deicing equipment, but it's never needed. You're up and down through the icing levels so quickly that none has time to build up. The other great bugbear of the early days was engine fires. Those big Wright turbo-compound pis-

ton engines in the Super Connies and Stratocruisers were just never very reliable. Propellers occasionally ran away and you couldn't get them feathered. And if a fire did start, high-octane certainly burned well. Jets now burn low-flash-point kerosene, and there's no vibration with them. They don't shake themselves to bits like the piston engines. In fact, they have vibration sensors in the main bearings, because if there is a hint of a wobble the bearing must be changed. In some jets, the altimeters have to have a mechanical device to keep tapping them so they don't stick. You never needed that with the old piston airplanes.

The captain has not always been a line pilot. He did a stint on the management side during World War II, helping to set up his company's new

Wherever airplanes gather, space is a
requisite—both to display the bulky beasts and to
accommodate their hordes of admirers. This is
a Paris Air Show, among whose entries were Douglas
DC-8 (below) with decor by Alexander Calder;
Falcon 20, a French business jet (bottom); Lockheed
L-1011 Tristar powered by Rolls Royce turbofans (r);
and F-14 Tomcat (opposite), U.S. Navy's
versatile fighter plane.

transoceanic division. Later he was seconded to a newly established Asian airline that had made a deal to buy his company's expertise and borrow some pilots to train its own. And he was a company check pilot for a while, riding shotgun with other captains of his company to check their performance. A check pilot is paid a premium on his salary, but frankly, he found the job too frustrating and bid to go back to line flying. And then came the jets.

The coming of jet propulsion doubled both the speed and the payload of transport aircraft. The extra power meant you could carry the extra fuel needed for extra range. When he started flying the Atlantic, New York to Paris in 1946 in Connies, it took 17 hours 20 minutes, including refueling stops at both Gander and Shannon. Now he can sometimes do it, if the winds are right, in five and a half hours (no stops, of course). There is even one version of the 747, the SP model, with such fantastic range that it can do New York-Tokyo nonstop, and fly right around the world with just two refueling stops.

The jet engine also provided, for the first time, the power to get up out of trouble. Unlike propellers, jet engines continue increasing in thrust as you accelerate (which is why jet climb speeds are so much higher than props). So if you should have to shut one down after takeoff, you can go right on up on the power of the other three. The drag of a windmilling jet engine is only a few hundred pounds, whereas that of a windmilling prop in the old Connie was more than three thousand pounds, at least until you got it feathered, and that really took a lot of rudder to fight and keep the airplane straight. On the 747, the powered controls are almost feather-light—a good bit lighter, in fact, than the manual ones on the much smaller 707. And a 707 really *is* much smaller: a 747's fuel, when fully loaded, weighs more than some 707s loaded for takeoff. In a 747 even a little guy can hold the rudder needed against a failed outboard engine without trim, while he

climbs to flaps-up altitude. That is *something*.

Airplanes now, the captain says, are so highly automated that they thrive on maintenance rather than piloting. (A 747 gets more than 10,000 man-hours of maintenance every year.) He learned to fly the 747 in just eleven days of ground school and less than four hours of dual in the actual airplane; it took him—and everyone else—three or four times as long to master the old 707. Mind you, there has been a small revolution in training methods. They taught him every last nut and bolt in the Connies—he can still remember the tailskid strut pressure for that airplane—even though a lot of the information was useless to him. The theme of 747 school is, if you can't do anything about something from your seat in the cockpit, forget it.

But he still finds it difficult to comprehend the whalelike size of the airplane. He knows that its span is several times longer than the Wright brothers' first flight, and that its fuel load is so enormous that it would keep a DC-3 flying for twenty-two days, or a Cessna 150 trainer for more than a year, continuously. It does sometimes bother him to think that he is burning up the earth's limited fossil fuel resources at such a rate. Hydrogen, liquid hydrogen, he thinks, is what airplanes will burn when the petroleum is all gone. Hydrogen made by electrolyzing plain water, using electricity generated by nuclear power or the sun's heat. Boeing, on the ball as ever, has done some studies and published drawings of a hydrogen-powered 747, with huge liquid hydrogen tanks mounted on the wings. Knowing Boeing, it would work.

Advancement for airline crews is strictly a numbers game. Promotion is by seniority in the company, not, as in most occupations, by merit. Since everyone has to pass the same check rides, merit is assumed to depend strictly on time served. Thus, when the line got its first six 747s and asked for volunteers to fly them, the captains' jobs went to the 245 graybeards, the six pilots with the most seniority. The copilot, much further down the seniority list but flying as a 707 captain, learned that he might get a 747 first officer's job if he was prepared to relinquish his captain's status for a year or two. After a good deal of soul-searching—it had taken him twenty-five years of flying to get into a 707's left-hand seat—he bid for the new bird and got it. Indeed, all the 747 first officers' jobs went to men who were then 707 captains. The flight engineers' jobs in the Jumbos went to the top 707 copilots and a few 727 captains who had bid for them. The result was crews of three men who averaged a total of seventy years in aviation among them, and seventy thousand hours of flying time.

Yes, the captain would like to fly SSTs, but he's sure he'll be retired before the U.S. airlines get them. The Anglo-French Concorde is a great airplane, but it is a step backward in productivity: it takes too many man-hours to manufacture and burns up too much fuel carrying too few passengers. The U.S. SST, when it comes, will be a quantum jump forward—probably a Mach 3 titanium airplane carrying 250 passengers against the Concorde's Mach 2.2 and 120 passengers (and on some routes they can't even carry the full 120). It might be a swing-wing airplane like the F-111, as was Boeing's first SST design. He thinks the engineering and weight problems that led Boeing eventually to reject the swing-wing concept will be worked out by the time the U.S. supersonic is introduced. One way or another, he expects, it'll have a tailplane, unlike the Concorde. With the trimming power of a tailplane you can have wing flaps (the Concorde has none) to get your takeoff and landing speeds down. For all its size, the 747 takes off and lands in the same speed range as a 707, and perhaps that—approaching 200 mph at the moment of takeoff—is as fast as any heavy airplane ought to be going on the ground. Maybe two miles of runway is as much as any transport airplane should need.

They let him fly the Concorde simulator at Toulouse in France once, and gave him a ride as a passenger in the airplane itself. What was it like? It was an airplane ride, much like any other; but it certainly was fast, and that is in the end what air transportation is selling—speed. Back in the twenties the designer of the Ford Trimotor, the first real airliner, predicted that people would one day fly from Detroit to New York in three and a half hours, and they laughed at him. Now the Concorde does Washington-London in about that.

In the cockpit of the 747, instead of the unceasing thunder the engines are sending back behind us, we hear only a high-pitched whine and the rush of air around the cockpit. The instruments are lit with a soft multicolored glow, amber and blue; a patch of blue light on the cockpit floor is no more than moonlight creeping down through the windshield. Curtains of stars hang outside the windows, seeming to be all around, even underneath us. One bright unblinking planet on the nose gives the illusion of being the light of another airplane, but it comes no closer than ten million miles.

Dawn begins ahead and to the left, as the sun, which disappeared earlier behind our shoulders, begins to push up in front of us. In the fading darkness the stars vanish, and the horizon begins to glow, first dark blue, then mauve, then green, shading to yellow and gold. And then the bright sun appears, very fast, for our own speed has doubled the natural speed of sunrise. As light returns to the sky, far below us we can see that the ocean wind has stretched little lines of cloud across the sea. For the first time since takeoff there is a visual sense of tremendous height and speed.

Soon our approaching landfall at Eagle Island off the coast of Ireland is indicated by life returning to the VOR and DME indicators on the instrument panel. Ireland rushes by below, quite unseen under a carpet of silky cloud, and we receive clearance to begin our descent while we are still more than a hundred miles from our destination. Out come the check lists again on the first officer's knee, and the litany of call and response begins once more, even before the descent is started.

"Pressurization?"

"Set, all packs on."

"Humidifiers?"

"Off."

"Pressure altimeters?"

"Cross-checked."

"Landing data?"

"Indexed and bugged."

"HSIs?"

"Radio mode."

"Transfer switches?"

"Guards closed."

And so it goes.

The radio beacons, now that we are again over land, come up and go by quick as fence posts. And the English air-traffic controllers on the R/T are somehow tremendously polite as well as efficient.

The cloud deck clears, revealing emerald-green fields, winding rivers, and tightly clustered towns in the morning sun. London's Heathrow Airport slides by on our left; we must go a fair way beyond it before being "vectored" by the radar approach controller to turn back and fit in with other traffic with four or five miles separation. Rather than letting the automatics land the airplane this time, the captain has invited his first officer to make the approach and touch-down, hand-flying the airplane, just for practice. "If you don't bounce it a couple of times I won't let you do it again," he says with mock seriousness. "I don't want you showing me up." As they roll out onto final landing heading, the Glide Slope needles on the panel come alive and start to signal a slow descent. Two jets exactly ahead can clearly be seen in the clean morning sky. "Leave gear and flaps till ten miles DME," says the captain. "No need to make any more noise than we have to." When they are down, both gear and flaps increase the airplane's drag substantially, and power and noise have to be increased to counteract this and to maintain approach speed. Though much of the original fuel load is now gone, the 747 is still a tremendously heavy airplane, and it needs a lot of power to accelerate it should it fall below target speed on finals.

The approaching runway is clearly in sight before flaps and gear go down. The flaps are enormous triple-slotted Venetian blinds wound back as well as down by massive worm drives; they give the 747 its low landing speed. As the urban sprawl of London's suburbs slides up toward the airplane, the captain calls height readouts from the radio altimeter, normally the first officer's job. Three hundred feet; 200 over the approach lights' crossbar; 100 over the huge painted numbers at the start of the paved runway; close the throttles at 60; flare at 30. *Aircraft* height, that is, for the cockpit is still 50 feet up at the moment of touch-down. Lower the nose, and apply the brakes and select reverse thrust. The turn-off from the runway is made at a speed that seems absurdly slow, but 15 knots is about the maximum for a 90-degree turn, or you risk scrubbing the nosewheel tires or even peeling them off the wheel hubs.

The passengers are out of the airport long before the crew, who have reams of paperwork to take care of before they are finished. It is more than an hour before they are on their way in the minibus to the downtown hotel. And they are weary—even after a lifetime of practice, you cannot miss a night's sleep and not be tired—but the airplane isn't. It is due out on a return flight to New York in a couple of hours. "Those big rascals just don't get tired like we do," says the captain by way of explanation. "Two Atlantic crossings a day is routine for jets; three is not unknown, and two and a half not uncommon. You can't leave a twenty-two-million sitting around gathering moss—it's got to earn its keep."

Flight crews' maximum work and minimum rest times are closely defined by law. The captain wears two watches, one set to Coordinated Universal Time (formerly called Greenwich Mean Time, and used throughout the world for navigation) and the other to "tummy time." The latter shows that this bright morning is really a very late night for him.

"Sometimes," he says, "I wonder if driving away from the airport afterward isn't really the nicest part of flying. Do you realize I've crossed the Atlantic 749 times, and not once by sea? That I've spent twenty thousand hours—more than two solid years—flying?

"No, I don't mean it; it's been a marvelous life. I think I'm as content with my work as a man can be. There's nothing I've ever wanted to do with my life so much as fly, and I've been lucky enough to get to do it. How many people can say that, that their work is the one thing they've ever really wanted to do? I wouldn't change places with anybody."

247

Photo Credits

BB—Berl Brechner
DN—David Namias
IWM—Imperial War Museum, London
JG—James Gilbert

Cover: JG. Page 2: JG.

Chapter 1
10–11: DN. 14: BB (top, bottom right); DN (bottom left).
15: Michael Philip Manheim, Photo Researchers
(top left); BB (top right); Dave Burnett, Photo Researchers
(bottom). 16: Russ Kinne,
Photo Researchers. 17: DN. 18–19: DN. 20–21: DN.

Chapter 2
22–23: JG. 25: Science Museum, London. 26–27: JG.
29: JG (top); Author's Collection (bottom).
31: Musée de l'Air, Paris. 32–33: IWM. 35: JG (top);
IWM (bottom). 36–37: IWM. 39: IWM. 40: IWM. 42–43:
JG. 45: Ryan Aeronautical Co. (top); Sikorsky
Aircraft Corp. (bottom). 46: JG. 47: USAF (top left); JG (top
right, bottom). 48: IWM (top); JG (bottom). 50–51: Author's
Collection. 53: National Aeronautics and Space
Administration.

Chapter 3
54–55: JG. 59: JG. 61: Russ Kinne, Photo Researchers (top);
JG (bottom). 62–63: DN (top left); BB (top right);
JG (bottom). 66–67: DN (top); BB (left, bottom).
69: DN. 70–71: BB. 72–73: Mike Charles (top);
JG (bottom). 74: *Flying*. 75: Mike Charles.
76: *Flying*. 77: National Aeronautics and
Space Administration. 78–79: BB.
81: DN. 82: BB. 83: Harry Brocke.
84: Learjet/*Flying*. 86–87: Russell Munson. 88: DN.

Chapter 4
All pictures JG.

Chapter 5
All pictures JG.

Chapter 6
All pictures JG except 148: Author's Collection.

Chapter 7
150–151: JG. 153: David P. Esler. 154–155:
David P. Esler. 156–157: David P. Esler. 158: JG. 159:
JG. 161: JG. 162–163: JG (top left); all others David P.
Esler. 164–165: David P. Esler (left); JG (right). 166:
JG. 167: JG (top); David P. Esler (bottom). 168: JG.

Chapter 8
170–171: JG. 172–173: Ted Koston. 174: Ted Koston
(top); JG (bottom). 175: JG. 176: Ted Koston. 178–179:
JG. 180: JG.

Chapter 9
182–183: JG. 186–187: JG. 189: BB. 190–191: JG.
193: Reims/Cessna. 194–195: JG (top, bottom
left); BB (right). 196–197: JG.
198–199: JG. 200–201: Air Portraits.

Chapter 10
202–203: USAF. 205: General Dynamics. 206–207: USAF.
208–209: USAF. 210–211: JG. 212–213: JG.
214: USAF. 216: JG. 218–219:
JG. 220–221: Peter Stevenson (top); JG (bottom).
222–223: JG. 224: JG.

Chapter 11
226–227: JG. 230–231: Pan American World Airways
(left, top right); Tom Stack (bottom right). 232: *Flying*.
234–235: John Madeley, Tom Stack Assocs. 236–237: JG.
238–239: JG. 240–241: JG. 242–243: JG. 244–245: JG.

Index

249